Guided MATH Conferences

Author
Laney Sammons, M.L.S.
Foreword
Deborah Allen Wirth, M.Ed.

SHELL EDUCATION

Publishing Credits

Robin Erickson, *Production Director*; Lee Aucoin, *Creative Director*;
Timothy J. Bradley, *Illustration Manager*; Sara Johnson, M.S.Ed., *Editorial Director*;
Aubrie Nielsen, M.S.Ed., *Senior Editor*; Grace Alba Le, *Designer*;
Corinne Burton, M.A.Ed., *Publisher*

Image Credits

Cover, CEFutcher/iStockphoto; all other images Shutterstock

Shell Education

5301 Oceanus Drive
Huntington Beach, CA 92649-1030
http://www.shelleducation.com

ISBN 978-1-4258-1187-7

© 2014 Shell Educational Publishing, Inc.

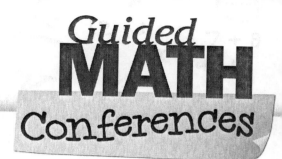

Table of Contents

FOREWORD

Thirteen years ago, I was leading twenty-five eager first graders through calendar math as part of our morning meeting. As students contributed a variety of ways to represent the calendar date, one student offered a multiplication number sentence. Knowing I had taught his older brother, I thought this was just a memorized fact he had learned from his sibling. Yet as I delved more deeply into his thinking, I found that not only did he clearly understand the concept of multiplication, but four other students in my first grade class had a firm grasp on this skill, as well! At the same time, several of my first graders were still grappling with one-to-one correspondence. The range of mathematical knowledge was vast. I knew I needed to change the way I instructed my students in mathematics. With the success my students and I experienced with Guided Reading, it became clear that I needed to utilize the parallel strategy of Guided Math.

The transition to a Guided Math classroom was fairly seamless, despite the lack of resources supporting the topic at the time. However, this was only possible because of the trials and tribulations I had worked through to implement Guided Reading. As I tried various structures for implementing the Guided Math strategy, one thing became very clear: I needed to become more intentional about finding out what kids knew mathematically. As Sammons's book clearly reverberates, this is best accomplished through Guided Math conferences.

Sammons's book engages the reader with an easy-to-follow format. Grade-level conferencing snapshots are peppered throughout the presentation of the six conference types: Compliment Conferences, Comprehension Conferences, Skill Conferences, Problem-Solving Conferences, Student Self-Assessment and Goal Setting Conferences, and Recheck Conferences. Sammons's delineation of conferencing types gives teachers a richer perspective with regard to the purpose of conferring. To offer increased insight, she also provides a conference structure so teachers may navigate intentionally through a conference rather than haphazardly muddling through potentially less meaningful exchanges.

Laney Sammons recognizes the challenges teachers face. She tackles these difficulties head-on and presents solutions that are viable in our ever-changing classrooms. She clearly outlines a structure that is practical, simple, and gives assessment *for* learning, rather than solely assessment *of* learning. This, I am convinced, is the hallmark of what we, as teachers, need to implement more readily. Sammons presents a myriad of ways in which Guided Math conferences accomplish this goal.

While I could cite a host of research supporting math conferencing, and I can offer compelling data from my own classroom—100 percent of my third-grade class scored "advanced" on our state's high stakes mathematics assessment—to try to persuade readers to pursue the strategies Sammons presents in this book, the practitioner in me knows that what matters most to educators holding this book right now is this testimony: *Guided Math conferences work!*

This is a must-have resource regardless of where you view yourself on the Guided Math strategy continuum. As a teacher with 29 years of experience and an experienced practitioner of the Guided Math strategy, I thought this book would primarily serve to validate what I already do in my classroom. But as I sat reading Sammons's book one weekend, I found myself eager to get back to my own classroom Monday morning to better incorporate the conferring techniques she outlines. With Sammons's fresh spin on the structure teachers can utilize when incorporating math conferences, I found myself motivated and excited to better meet the needs of my mathematicians. Her conferring framework gives immediate feedback to both the student and the teacher while encouraging both populations to think more deeply and critically. Her book now serves as a foundational framework from which I can become more effective with the Guided Math strategy I have incorporated for more than a decade. Without a doubt, you will be inspired by this compelling resource. Enjoy!

— Deborah Allen Wirth, M.Ed.
Guided Math Consultant, Bureau of Education & Research

ACKNOWLEDGEMENTS

First-time authors usually enjoy a high degree of support. When a family member decides to write a book, it is an exciting event. There is much enthusiasm from the immediate family. Others graciously step in to assume routine household tasks that previously have been done by the author. And then, the whole family shares in a sense of accomplishment when the book is finally published.

When authors choose to continue writing, however, the impact on their families is multiplied and ongoing. What begins as something novel and exciting becomes commonplace. It is only the most fortunate authors who continue to experience the same level of family support. For the ongoing support and enthusiasm of my family, I am so grateful. My love and thanks to my husband, Jack, my daughter, Sorrel, and her family, and my son, Lanier, and his wife, for their patience as I write.

I also wish to thank the many teachers and educational leaders with whom I am privileged to work. I learn so much from them and feel honored to work in their schools and districts—especially when I am invited into classrooms to work directly with their students.

And finally, I would like to express my appreciation to Sara Johnson and Aubrie Nielsen, my editors, and the staff at Shell Education who have contributed to the publication of my books and have allowed me to share my ideas with educators.

PREFACE

When Teachers and Students Talk about Learning

> *When teaching and learning are visible, there is a greater likelihood of students reaching higher levels of achievement.*
> (Hattie 2012b, 18)

What constitutes good teaching? For decades this has been much debated. Everyone, it seems, has an opinion. Yet there has been no consensus on this issue. Instead, much attention is directed at debating the merits of the latest trends in educational reform. Perhaps, rather than search for reforms, it may be more productive to focus on teaching practices that have been shown to have the greatest positive impact on learning, and then consider ways in which these can be carried out in our classrooms.

In his book *Visible Learning*, John Hattie (2009) reports on what he discovered from his study of more than 15 years of research on what works in schools. In his more recent tome, *Visible Learning for Teachers: Maximizing Impact on Learning*, he examines the attributes of teaching that have the most impact on learning (based on his research synthesis) and explores their implications for teachers. What struck me was his unsurprising—yet rarely articulated—conclusion that *teaching and learning must be visible*. As he puts it, what makes a difference are "…teachers seeing learning through the eyes of students and students seeing teaching as the key to their ongoing learning" (2012b, 14).

Hardly surprising! Rather than a one-size-fits-all script for teaching, student achievement is greatest when teachers strive to find out what their students know and are learning and then adjust their teaching to meet student needs, and when students assume the responsibility for both knowing what their learning goals are and for monitoring their progress toward meeting them.

When teaching literacy, conferring with my students about their reading and writing made their learning much more visible to me. These one-on-one conversations with my students gave me glimpses of their learning through "their eyes." Armed with this knowledge, I could then adjust my teaching to make it more effective.

Moreover, conference conversations also served to focus my students' attention on what I was teaching and why it was important to them. They began to more clearly recognize that the teaching points during the conference were crucial to their learning. Because they were expected to talk with me about their learning, they became more cognizant of it and began to assume a greater responsibility for monitoring it.

To me, there is an obvious connection between Hattie's conception of visible teaching and learning and the practice of teacher and students conferring. The benefits of these conferences are not limited to literacy. Good teaching is good teaching—across the content areas! The same conferring techniques that have been so effective for literacy are a crucial component of the Guided Math framework. These thoughts led me to write this book. And, assuming Hattie's findings are valid, the increased visibility of both teaching and learning which results from math conferences will lead to greater mathematical achievement for students.

CHAPTER 1

Conferring with Young Mathematicians

Just as students come in all shapes and sizes, with distinctive personalities, quirks, senses of humor, and sensitivities, they also come into our classrooms with unique background knowledge and instructional needs. Somehow we manage to adapt our instruction each year as these young learners enter our classrooms—not only adapt, but truly delight in the diversity that enriches the learning environment we so carefully construct.

Realistically, however, the vast differences in foundational knowledge and skills of students pose challenges—challenges that are intrinsic to the profession of teaching. How do we gain insights into the thinking of our students, discover what they know, what they can do, what misconceptions they have, what struggles they face, and what concerns they harbor? The intimate nature of small classes allows astute teachers to establish close relationships with their students in which they acquire something of a true measure of their students' learning strengths and needs through observation and discussions. Building on that measure, differentiation of instruction flows naturally.

With larger class sizes, the task of accurately assessing the complex and unique individual learning needs of our students is more difficult. Opportunities to closely observe their work are limited, as are in-depth student-teacher conversations during lessons, especially during whole-class lessons. In some ways, a classroom full of students is analogous to an orchestra. When twenty-five or thirty students are each playing their own scores, the blended sound of the whole orchestra makes it almost impossible to distinguish the sounds of individual instruments. Nuthall (2005, 919) found that unlike conductors who can sometimes pick out the sounds of individual musicians, teachers in such settings are "largely cut off from information about what individual students are learning" and "are forced to rely on secondary indicators such as the visible signs that students are motivated and interested."

Diagnostic tests, benchmark tests, and other paper-and-pencil assessments may give us some guidance, but they frequently fail to expose students' thinking or provide deeper insights into their background knowledge. Too often, these methods of assessment let us know only whether students were able to find or choose the correct answers—not necessarily whether they really *know how* to find the answer or whether they have any misconceptions. Paper-and-pencil assessments do not often shed much light on the depth of students' understanding of mathematics.

A Tale of Four Students

To illustrate the limitations of these assessments, consider the mathematics assessment results of four hypothetical students on a written multiple-choice test. Two of these students chose the correct answer for a question, while two others chose an incorrect answer. Since the assessment was not an open-response test, what does their teacher *really* know after grading this assessment? Did the students who chose the correct answer know how to solve the problem? In this case, one of these students chose the answer solely by chance. The other found the correct answer using incorrect reasoning. Based on the correct answers, one might assume that both students understood the concepts or skills being tested, while in fact, neither of them did.

On the other hand, consider the two students whose answers were incorrect. In this hypothetical, one of the two really understands the concepts involved and knows how to solve the problem, but made a simple error in computation. He may have been tired on the day of the test, or possibly lacked the motivation to fully work out the correct answer. Perhaps he was ill or troubled about something in his personal life. But the isolated assessment results give a false indication of his level of knowledge.

Finally, the last of these four students had absolutely no idea how to find the correct answer. But from the written test, how do we know that for certain? When a student's answer is incorrect, we need to know why. Where did the young learner go wrong? What gaps in knowledge or skills prevented the student from being able to find the answer? Meaningful, relevant instruction for students can only be planned and delivered when these questions are answered.

The confusion about the mathematical ability and learning needs of these four students is obvious even when considering their answers to just one question. The lack of precision in targeting instructional needs only increases when an assessment is composed of many questions. This kind of assessment may offer teachers a snapshot of the overall achievement of their students, but little specific guidance concerning the instructional needs of individual learners.

Open-response assessments yield a clearer picture of what students can and cannot do, as well as what they do and do not understand. Although the results from these assessments may inform our planning of instructional "next steps" for individual students based on their needs, our descriptive feedback to students is often delayed. And even with this assessment format, we are routinely in a position of having to guess exactly what our students' thoughts were as they worked—particularly with students whose ability to communicate mathematically is limited.

But what about assessing student understanding during class discussions? Won't students' oral responses provide us with more transparent evidence of their thinking? After all, most of us ask questions frequently during lessons to gauge student comprehension. Too often, however, even with the use of questioning, it is difficult to clearly discern what students are thinking. Marilyn Burns (2005, 27) shares her early experiences as a teacher:

> …*when students answered questions correctly I usually accepted their responses with a nod or comment of approval, rarely prodding them to explain their reasoning. When students were incorrect, however, I was more likely to probe further by asking, "Are you sure about that?" or "Why do you think that's right?" Follow-up prompts like these then became signals to the students that their response was not correct or acceptable.*

Even when using discovery learning, Burns found that she failed to probe students' levels of understanding when correct answers were given to her questions. She states, "I never really knew what students were thinking or whether their correct answers masked incorrect ideas. I only knew that they had given the answer I sought" (2005, 27).

When we ask probing questions of our students during class discussions, we may indeed learn what one or two, perhaps even three, students are thinking, but it is folly to assume that those thoughts accurately represent the thinking of the entire class. Yet without more in-depth knowledge, we are bound to fail in our mission of meeting the needs of all our students.

A Glimpse into Student Thinking

Literacy teachers have long been familiar with student-teacher conferences. Conferring with young learners is integral to both reading and writing workshop (Anderson 2000; Calkins 2000; Calkins, Hartman, and White 2005; Graves 2003; Serravallo and Goldberg 2007). Young learners share their thinking about their personal reading or writing in one-on-one conversations with their teachers. Not only do teachers discover much more about their students' capabilities and next steps in learning, but close bonds between students and teachers are also formed.

The benefits of student-teacher conferences are not exclusive to literacy instruction. Sound instructional practices span the content areas. Serravallo and Goldberg (2007, 1) describe their beliefs about reading as follows:

- Reading is the act of constructing meaning.
- Reading is a process.
- Reading is deeply personal and, therefore, varies from reader to reader.

Indeed, mathematical work is very similar. Mathematicians engage in constructing meaning from their prior and current mathematical experiences to understand new concepts and solve problems. They participate in a process as they determine meaning, make connections to other areas of mathematics, and draw upon their background knowledge. They employ strategies they have acquired to deepen their knowledge and to find solutions to problems. And finally, understanding mathematics is deeply personal and varies from mathematician to mathematician. That is not to say that there are not constant mathematical principles, but an individual's perspectives, modes of learning, and previous experiences combine to make his or her approach to comprehending mathematical concepts and problem solving entirely unique.

Furthermore, with these beliefs about reading in mind, Serravallo and Goldberg (2007, 7–8) conclude that effective reading instruction should:

- match the individual reader;
- teach toward independence;
- explicitly teach strategies;
- value time to experience reading; and
- follow predictable structures and routines.

These beliefs about reading instruction may be easily adapted to describe effective mathematics instruction. As mathematics teachers, we are most effective when our instruction matches our individual learners, we teach toward independence, we explicitly teach strategies, we value time for our students to explore challenging mathematical problems and concepts, and we establish and then maintain predictable structures and routines.

Because of their beliefs about reading and reading instruction, Serravallo and Goldberg (2007) understand the importance of one-on-one reading conferences, in which instruction is targeted specifically to student strengths, nudging young learners to the edge of what they are just beginning to be able to do, and supporting them as they begin to independently apply new strategies they have learned. The intimate nature of conferences allows teachers to know their students so well that their teaching points for each student are at an instructional level that is most appropriate to the student's immediate needs and current developmental phase.

Thus, in many ways, conferring with students is the heart and soul of teaching (Sammons 2010). According to Calkins, Hartman, and White (2005, 6), "It gives us an endless resource of teaching wisdom, an endless source of accountability, a system of checks and balances. And, it gives us laughter and human connection—the understanding of our children that gives spirit to our teaching." Conferring gives teachers true glimpses into the thinking of their students—whether it be thinking about reading, writing, or mathematics. With these glimpses, our instruction becomes more focused and powerful.

What Are Guided Math Conferences?

The Guided Math framework (Sammons 2010) is composed of seven components: establishing a classroom environment of numeracy, math warm-ups, whole-group instruction, small-group instruction, math workshop, math conferences, and assessment. (See Appendix A for a description of each component of the framework.) Together, these instructional tools offer teachers a manageable means of identifying and then meeting the mathematical needs of their students. One-on-one math conferences are valuable for accurate assessment of student strengths and needs, and for targeting individual needs through timely feedback and brief specific instruction—thus, they are an important support for the other six components. Even in classrooms where the Guided Math framework is not being used, conferring one-on-one with students gives teachers rich insights into their students' mathematical thinking as well as opportunities to provide effective, targeted instruction and offer constructive feedback.

Math conferences are one-on-one *conversations* with students about their mathematics work, as one mathematician talking with another. Literacy educators have written much about successfully conferring with students. Murray (2004, 148) emphasizes the importance of maintaining a conversational tone during conferences. "They are not mini-lectures but the working talk of fellow writers sharing their experience with the writing process. At times, of course, they will be teacher and student, master and apprentice, if you want, but most of the time they will be remarkably close to peers…."

Like conversations, conferences between teachers and students include these characteristics (Anderson 2000):

- Conferences have a purpose.
- Conferences have a predictable structure.
- Lines of thinking are pursued with students.

- Teachers and students each have conversational roles.

- Students are shown that teachers care about them.

We show our genuine interest in the work of our students when we confer. Sitting side-by-side and shoulder-to-shoulder with our students, we dig deeper so we can meet their unique, individual needs. In doing so, we support them as they begin to apply what they are learning in both large-group and small-group lessons (Miller 2008). The thoughtful conversations we create get to the core of their thinking and then prompt them to consider what they are doing from other angles or with more depth.

Math conferences are a time for students to share their mathematical thinking with their teachers. In doing so, they learn not only to organize and express their mathematical ideas cogently, but also to continually reassess the validity of their reasoning. Moreover, these mathematical conversations support the learning of new concepts and strategies by requiring that students focus on representing their work, both verbally and with diagrams, models, or symbols so that it can be clearly understood by others. All of these are essential aspects of mathematical practice (Common Core State Standards Initiative 2010; National Council of Teachers of Mathematics 2000).

Math conferences are also a way of extending and deepening the numeracy of our students. Allington (2012) describes the ability to go beyond word calling, simple recall, and recitation when reading as *thoughtful literacy*. It requires a reader to engage with the ideas in a text, challenge them, and then reflect on them. When conferring with students, we foster that same kind of understanding of mathematics—*thoughtful numeracy*, the mathematical counterpart to thoughtful literacy. We help students develop the mathematical skills they need to cope with the practical demands of everyday life (Steen 1990). With that goal in mind, these conversations between students and teachers serve to increase the capacity of young mathematicians to effectively engage in mathematical thinking and problem solving, critically consider the mathematical data and the reasoning of others, and clearly communicate their own mathematical thinking, so that they will be able to successfully apply the knowledge, skills, and strategies they have acquired to new situations and problems they encounter throughout their lives (Saskatchewan Ministry of Education 2009).

Guided Math Conferences, Math Interviews, and Small-Group Instruction

Instructional strategies for teachers abound. As we focus on conferring with students about mathematics, it is important to distinguish between math conferences, math interviews, and small-group instruction. All three share important characteristics and have a place in the Guided Math framework, but each is unique in many ways. Although all of these instructional elements afford teachers opportunities to build relationships with their students, encourage mathematical communication, and assess student understanding, they vary in the amount of time required, the participants involved, the focus of the conversations, and their primary functions. A summary of the comparisons that follow can be found in Figure 1.1.

Guided Math Conferences vs. Math Interviews

Looking first at math conferences and math interviews, both are one-on-one conversations between a teacher and a student. They are valuable forms of assessment during which teachers learn more about the mathematical understanding and capability of their students. They both uncover students' misconceptions and gaps in understanding that may not be apparent when relying only on the written work of students. According to Burns (2010, 19), her experiences with student interviews were "revealing and sometimes astonishing," exposing "the fragile conceptual base of [students'] understanding that their teacher had no way of knowing from the context of the classroom lesson." Thus, the information from both conferences and interviews can serve to guide instructional decisions. In addition, both techniques are powerful ways in which teachers can connect more deeply with their students as students share their mathematical thinking with the teacher.

In spite of the similarities, however, there are some distinct differences between these two instructional practices. Although math interviews are not a part of the Guided Math framework, they can be used for diagnostic assessment in a Guided Math classroom. In math interviews, the main focus of conversation is a given task proposed by the teacher during the interview based on a "big idea," with specific questions to determine the degree of student understanding and to expose any existing misconceptions. The primary function of the interview is that of assessment to inform later instruction (Moon and Schulman 1995). No feedback is given to students during the interview. Instruction based on strengths and needs that are uncovered during the interview is delivered at a later time. The length of time required for an interview is usually about ten to fifteen minutes.

In contrast, math conferences are usually about five minutes in length. The focus of the discussion between the student and teacher is the mathematics with which the student is currently working. With this conversation, the teacher is conducting research to discover both student strengths and needs. The teacher uses this information to provide immediate, specific feedback and to decide on an appropriate instructional "next step" for the student. Then, within the conference itself, the next step is taught. As such, the major functions of the conference—assessment, feedback, and instruction—are entwined.

Both math conferences and math interviews offer teachers excellent ways to probe student thinking—going deeper than is possible in large-group or even small-group instruction. Because of the brief nature of math conferences, however, teachers are able to conduct them more frequently and more spontaneously throughout the school day. This flexibility is a considerable advantage to teachers whose instructional time is already strained by ever-increasing demands.

Math Conferences vs. Small-Group Instruction

Both math conferences and small-group instruction are essential components of the Guided Math framework (Sammons 2010, Sammons 2012). Although they are both powerful tools for teachers, the differences between the two are many—in the participants involved, the durations, the focuses, and the functions.

The small-group instructional format involves the teacher meeting with groups of two to six students with similar instructional needs for fifteen to twenty minutes (Fountas and Pinnell 1996 and 2001; Sammons 2010). Working in their "zone of proximal development," learners supported by the teacher expand and extend their mathematical understanding and capabilities (Vygotsky 1978). Teachers present brief mini-lessons about concepts or strategies to be learned and then actively engage students in practice requiring the young mathematicians to stretch just beyond what they can do successfully on their own.

Throughout the small-group lesson, teachers provide just enough support to move their students to a higher level of independent mathematical competence. The primary focus of the small-group format is the lesson itself, planned to meet the needs of the members of the group. As the lesson is presented, teachers are also able to informally assess student learning and to offer feedback based on the student work observed and the conversations that arise. In addition, small-group lessons "nurture joy, rigor, and empowerment" and inform other components of mathematics instruction (Wedekind 2011, 26), all of which are essential to effective implementation of the Guided Math framework.

While the worth of small-group instruction cannot be overstated, the intimacy of a one-on-one conference is absent in this setting. When conferring with students, teachers meet with just *one* student at a time. They ask students to tell them about the mathematics with which they are currently working, informally assessing student understanding and skills to identify both strengths and needs. As individual needs are discovered, teachers try to determine the best immediate instructional "next step" for these students. Which of the student needs noticed by the teacher can be most effectively targeted in the brief conference? And then, how can that teaching point best be conveyed to the student?

An integral part of conferences is immediate and specific feedback for students. Teachers share with them both something they have done well, in the form of an authentic compliment, and what they can do to make their work better or to increase their understanding. A learning goal, usually an incremental one given the brief nature of a conference, may be suggested. Along with the timely feedback, teachers might share a new strategy, correct a misconception, model mathematical communication, or demonstrate a process to address the specific teaching points for the young learners with whom they are conferring. Thus, within the Guided Math conference structure, teachers assess, give feedback, and teach a logical "next step."

The opportunities to give individual students both specific feedback on their mathematical work *and* instruction that targets the unique teaching points of individual students during a small-group meeting are rare. So in addition to the differences in the number of students involved, the time required, and the focus of small-group instruction and math conferences, a major advantage of the conference format is the ability to promote specific individual learning. A challenge, however, is creatively finding the time to engage in these mathematical conversations with students on a regular basis.

Figure 1.1 Comparisons of Math Conferences, Math Interviews, and Small-Group Instruction

Format	Time	Participants	Focus	Function
Math Conference	About 5 minutes	Teacher and 1 student	Student's current work	• Assessment • Feedback • Individual Instruction (Teaching Point)
Math Interview	About 10–15 minutes	Teacher and 1 student	Instructional task introduced by teacher	Assessment
Small-group Instruction	About 15–20 minutes	Teacher and 2–6 students	Group lesson (based on identified needs of the group)	• Group instruction (Focus lesson) • Assessment • Feedback

Snapshot of a Math Interview

Terrence has recently enrolled in a second-grade class. Although his teacher received Terrence's grades from his previous school, she hopes to gain a little more insight into her new student's understanding of place value by conducting a math interview. As her other students are engaged in independent work, she sits down one-on-one with Terrence.

Teacher: *Terrence, I am so happy you have joined our class. As a teacher, I really try to discover what my students think when they work as mathematicians, so that I will know how to best help them learn math. Thank you for being willing to talk with me for a few minutes.*

Terrence: *That's okay. I'm not great at math though.*

Teacher: *Well, don't worry about what we are doing today. Just answer the questions I ask as best you can.*

Terrence: *Okay.*

The teacher places two sticks of ten linking cubes each and 4 individual cubes on the table. On a white board, she writes 24.

Teacher: *Please read this number.*

Terrence: *Twenty-four.*

The teacher points to the 4.

Teacher: *Please show me what this part of the number represents using the cubes.*

Terrence gathers together the four individual cubes.

Teacher: *Tell me why you chose those cubes.*

Terrence: *It's a four. So I got the four cubes. One, two, three, four.*

The teacher then points to the two.

Teacher: *Now, show me what this part of the number represents.*

Terrence chooses the two sticks of ten linking cubes.

Terrence: *That's easy! Twenty is two tens.*

The teacher removes the cubes and replaces them with three sticks of ten linking cubes and fifteen individual cubes. She writes the number 45.

Teacher: *Please read this number.*

Terrence: *Forty-five.*

The teacher points to the 5.

Teacher: *Please show me what this part of the number represents using the cubes.*

Terrence chooses five individual cubes.

Terrence: *Five! Five cubes—One, two, three, four, five.*

The teacher points to the 4.

Teacher: *Please show me what this part of the number represents using the cubes.*

Terrence looks confused and stares at the cubes.

Terrence: *Is this a trick question?*

Teacher: *Why do you think that?*

Terrence: *Because you only put out three tens! I can't show you four tens when there are only three of them.*

Teacher: *I see. Let's talk about this idea on another day because it's time to stop our interview for today. Thank you for working with me on these numbers.*

The teacher discovered that Terrence can correctly read the double-digit number she wrote. He has a beginning knowledge of place value, but needs more work with ones and tens to understand that ten ones has the same value as a ten.

Because the interview was intended as an assessment of Terrence's understanding of place value to inform future instruction, the teacher gave Terrence no feedback on his responses, nor did she correct his misconceptions. The information the teacher obtained will be used to place Terrence in a small group of students with similar instructional needs.

During those lessons, his misconceptions will be addressed and he will receive more focused feedback on his work.

Snapshot of a Small-Group Lesson

The following lesson is for a fourth-grade class working on a unit on measurement and geometry.

Connection:

We have been working with angles—identifying types and finding their measures. We have also been examining two-dimensional shapes and their attributes.

Teaching Point:

Today we are going to explore the sum of the interior angles of convex polygons.

Active Engagement:

Students work in pairs to measure the interior angles of triangles and quadrilaterals, then share their results.

What is the relationship between the sum of the interior angles of a triangle and a quadrilateral? Is there any way you could have figured that out without measuring each angle of the quadrilateral? (Dividing the quadrilateral into two triangles)

Let's try a pentagon. Divide it into the fewest possible triangles and compute the angle sum. Measure to confirm. What pattern do you see? Do you think that pattern will always occur?

Link:

You have been thinking like mathematicians—looking for patterns in your work and making conjectures. Remember to use what you know about the sums of the interior angles of convex polygons when you are trying to solve problems. Share with the group something mathematical you have learned or thought about in this lesson.

The focus of the lesson is leading students to discover the relationships between the interior angles of polygons. The teacher observes the work of the students closely and listens to their talk as they work. At times, the teacher questions students to focus their attention, have them justify their reasoning, prompt them to think more deeply, or encourage their mathematical communication. During these interactions, the teacher provides appropriate feedback to students based on their work and discussions.

Snapshot of a Math Conference

Josephina is a seventh-grade English language learner student who struggles with problem solving, particularly with determining which operation should be used to find a solution. With this math conference, the teacher hopes to gain some ideas about how to help her become more proficient.

Josephina is working to solve the problem shown below. The teacher notices that she has underlined the word *all*.

Task:

A dress was on sale for 50% off its original price of $40. Later, the store took another 25% off that price. How much does the dress cost now after all the discounts?

Teacher: *Josephina, what are you working on?*

Josephina: *I'm trying to find the answer to this problem. It's kind-of hard. I keep reading it—I know you are supposed to do that. That's the first thing you should do.*

Teacher: *So as you read this problem, what are you thinking?*

Josephina: *I'm really thinking about the words in the problem. They can help me know what to do. I'm trying to remember what my teacher last year told me. She said there were certain words that tell you what to do to find the answer. I found a word that I think she talked about— "all"—and underlined it. I think that means I have to add things up to find the answer.*

Teacher: *Why do you think that means you should add to find the answer?*

Josephina: *Because there are certain words that tell you what to do. I think "in all" means add. But I'm not really sure how to add this up.*

The teacher realizes that an earlier teacher of Josephina's was probably aware of her difficulties understanding English and had encouraged her to turn to "key words" rather than teaching her to work to truly understand problems she was attempting to solve.

Teacher: *Josephina, as a mathematician, you know how important it is to first read and understand the problem you are trying to solve. You are working on the first step in problem solving! Let me share something with you that I have learned. When I am faced with solving a problem, I have to do more than just look for certain words. I have to really try to see in my mind what is happening, so I can figure out a way to solve it. Just because this problem has the word "all" in it, it doesn't necessarily mean that we add to solve it.*

Let me tell you what I do sometimes when I read a problem. I try to think about and "see" in my mind what is happening. So when I read that there is a sale, I know that I can buy something for less than what it used to cost. The store takes something off the price— it costs less to buy it when it is on sale. Then, I think about what mathematical operation I should use to show that something is taken off or away from the price. What do you think?

Josephina: *I think you subtract. Don't you?*

Teacher: *That's right, Josephina. When we read carefully and really tried to think about what happened in the problem instead of looking for certain words, we were able to figure out what to do to solve the problem. Do you remember how to find a percent?*

Josephina: *Yes. I just couldn't figure out what to do with it.*

Teacher: *Josephina, can you tell me in your own words what we did to figure that out?*

Josephina: *Well, for one thing, I didn't just look for those words like from last year.*

Teacher: *Well, what did we do?*

Josephina: *I put it in my head. I mean, when I read it, I made sort of a movie to see what was happening. Then, when I thought I knew what was happening, I knew I had to subtract. That dress wasn't going to cost so much.*

Teacher: *Now you are thinking just like a mathematician! Whenever you are trying to solve a word problem, do just what we did with this one. Try to see what is happening—just like making a movie in your head.*

Through questioning and listening to Josephina's responses during this conference, the teacher discovered that Josephina was trying to use key words to determine which operation to use when problem solving. Since Josephina knew it was important to read the problem carefully before deciding on a strategy for solving the problem, the teacher was able to give her an authentic compliment letting her know what she was doing well. The teaching point of the conference was the use of visualization of the problem rather than searching for a "key word" to decide which operation was required. Josephina was able to explain the teaching point in her own words. The teacher then reminded Josephina that this was something she should always do when solving problems. The teacher will closely observe Josephina's future work to be sure that she is making use of this strategy and reinforce it, if necessary.

The Structure of a Guided Math Conference

Most conversations we have with others have predictable structures. We are comfortable following patterns of verbal give-and-take that we experience day in and day out in our formal and informal relationships with those around us. We know how they begin, the various parts, the transitions from speaker to speaker or topic to topic, and then we know how they are brought to a close (Anderson 2000). Consider the most basic of greetings: *Hi, how are you doing?* The typical response follows: *I'm doing fine, thanks. And you?* The greeter responds, *I'm fine, thanks.* Who does not know this pattern? We all know how to initiate it, to respond, and where it leads. Whenever we engage in a conversation, we draw upon an unconscious knowledge about how to talk with others that we have been absorbing since birth. The predictable structure we know so well allows us to effortlessly begin our conversations and enables conversation to flow smoothly.

As with everyday banter, when math conferences have an established structure, the predictability allays anxiety, allows for the easy exchange of ideas, and leads to more productive discussions. The traditional teacher-student pattern of discourse in the classroom is question-response-evaluation. In *Choice Words: How Our Language Affects Children's Learning,* Peter Johnston (2004, 6) warns, "There is always an implicit invitation to participate in a particular kind of activity or conversation. We cannot persistently ask questions of children without becoming one-who-asks-questions and placing children in the position of the one-who-answers-questions." In contrast, the structure of a math conference is one that is more common to interactions between individuals sharing a similar interest. The teacher is simply someone who has genuine interest in the work of the student with whom he or she is talking and who is willing to share strategies for making that work even better.

Calkins, Hartman, and White (2005) describe a consistent and predictable architecture for writing conferences that can be adapted effectively for Guided Math conferences. The specific steps for these conferences are *research*, *decide*, *teach,* and *link*. The conference framework guides teachers as they confer, so they can discover what their students are thinking mathematically and then identify what to do to help them progress in both their understanding and skill. Following a structure gives purpose to what otherwise may be chatting without focus (Sammons 2010). The structure of a Guided Math conference will be elaborated upon in Chapter Three. Figure 1.2 provides an overview of this structure.

Figure 1.2 The Structure of a Guided Math Conference

The Structure of a Guided Math Conference

Research Student Understanding and Skills

- Observe the work of the student.
- Listen carefully as the student responds to questions about his or her work to understand what he or she is trying to do as a mathematician.
- Probe to glean more about the student's intentions, comprehension of relevant concepts, and mathematical capability.
- The student does most of the talking during this part of the conference.

Decide What Is Needed

- Weigh the validity of the student's current strategies and processes. Determine what should be the student's next step in learning. Decide on a specific teaching point and how you will teach it.
- Name specifically what the student has done well as a mathematician with an authentic compliment, linking it directly to the language of the standards, and remind him or her to continue to do this in future work.

Teach to Student Needs

- Use demonstration, guided practice, or explicit telling and showing to correct or extend a student's understanding and ability to successfully complete the task.
- Have the student briefly practice what was taught and explain what she or he has learned to ensure initial understanding.

Link to the Future

- Name what the student has done as a mathematician and remind him or her to do this often in the future.
- Have the student share a reflection on the mathematics learned.

(Adapted from Calkins, Hartman, and White 2005)

Kinds of Guided Math Conferences

While the conference content and teaching point will be largely determined during the research phase of the math conference, teachers often enter into conferences with preconceived areas of focus based on students' prior mathematical work. Experienced teachers know that even in these situations, much is to be gained by encouraging students to share their thinking and by listening carefully before selecting a teaching point. Nevertheless, Guided Math conferences typically fall into these categories. Each of these kinds of math conferences will be discussed in Chapter Four.

- **Compliment Conferences:** Teachers use these conferences to motivate young mathematicians or to lift the spirits of discouraged learners.

- **Comprehension Conferences:** The focus of these conferences is on assessing and then extending the degree of student comprehension of mathematical concepts.

- **Skill Conferences:** The aim of these conferences is assessing and then extending the skills of students, including both process and computation skills.

- **Problem-Solving Conferences:** These conferences are used to explore the problem-solving strategies being applied by students and then to strengthen their toolbox of strategies, if needed.

- **Self-Assessment and Goal-Setting Conferences:** Together, students and teachers review progress toward meeting learning targets and establish learning goals.

- **Recheck Conferences:** Teachers use these conferences when they want to see if students are using what they learned during earlier conferences.

Chapter Summary

Of utmost importance to the teaching of mathematics is having an awareness of precisely what students know and can do. One-on-one Guided Math conferences allow teachers not only to acquire specific information about the mathematical proficiency of individual students, but also to teach these students their "next steps" in learning within a highly supportive setting. Students receive immediate and very specific feedback about their work as a part of these discussions. As such, the Guided Math conference format is a way of assessing, teaching, and providing feedback.

Additionally, one of the most positive consequences of conferring is the nurturing of the student-teacher relationship. Students find they can talk freely about their work, sharing questions or confusions, in a nonthreatening setting. The structured nature of the conference ensures that the conversation is focused on math, holding students accountable and teaching them how to communicate their mathematical thinking.

REVIEW AND REFLECT

1. How often are you able to engage your students in one-on-one conversations about their mathematical thinking?

2. What do you think is the most important benefit of math conferences? What are the greatest hurdles to implementing math conferences in your classroom? How could you overcome these hurdles?

3. Think of a student in your class who is struggling with a mathematical concept or skill. What would you like to know about his or her mathematical thinking? What questions would you ask if you decide to confer with this student?

CHAPTER 2

The Value of Guided Math Conferences

Throughout the world of education, there is a growing demand for more rigorous mathematical instruction and learning. The performance of United States students on international mathematics assessments has spurred the public's concern over student achievement (National Center for Education Statistics 2010; Baldi et al. 2007). As teachers, we are re-examining traditional models for teaching mathematics and refining our teaching practices to reflect an emphasis on greater mathematical conceptual comprehension, computational fluency, and problem-solving capability for young learners. These aspects of learning demand more from students—they must learn to think critically about the discipline of mathematics.

Increasing Depth and Rigor

As states adopt ever more rigorous mathematics standards, including the Common Core State Standards (Common Core State Standards Initiative 2010), our students are now expected to develop a much deeper understanding of mathematical concepts. To meet these standards, our students must be able to draw upon a comprehensive understanding of math, to reason mathematically, and then to apply what they know to solve complex and often unfamiliar problems. Being proficient in math now requires considerably more than just following memorized procedures.

Thinking Critically about Mathematics

The National Council for Critical Thinking Foundation's website (Scriven and Paul, under "Summary") defines critical thinking as "skillfully conceptualizing, applying, analyzing, synthesizing, and/or evaluating information gathered from, or generated by, observation, experience, reflection, reasoning, or communication, as a guide to belief and action." It is a mode of thinking in which one skillfully uses the structures of thinking to improve the quality of thought. When thinking critically, we:

- consider questions and problems;

- gather, assess, and interpret relevant information to form conclusions;

- think open-mindedly, recognizing assumptions, implications, and consequences;

- communicate effectively with others to find solutions to complex problems (Scriven and Paul).

It is just this kind of rigorous thinking that the new Standards for Mathematical Practice promote (Common Core State Standards Initiative 2010), and it does not often come easily to young learners, but the nature of math conferences is ideal for leading students to think more critically. As teachers confer with students about mathematical work, they are able to share their own insights, strategies, and curiosity about the discipline. Careful questioning can challenge facile thinking and encourage young mathematicians to dig deeper as they are asked to examine more intently their assumptions about mathematics. It is ongoing experience with this kind of mathematical reasoning that supports students' higher-order thinking skills (Conklin 2012).

When conferring with students, we model how to express mathematical ideas clearly and how to make mathematical connections by drawing upon background knowledge. We share our enthusiasm for the discipline during these verbal exchanges with students. Perhaps of even greater importance is the expectation we convey during math conferences that our students also actively participate in the dialogue by explaining their thinking and wonderings. Certainly, the promotion of rigor and critical thinking in mathematics instruction is possible with other teaching formats, but the impact is magnified with the focus and intensity of these one-on-one math conferences between students and teachers.

The Strands of Mathematical Proficiency

The National Research Council (2001) report *Adding It Up: Helping Children Learn Mathematics* identified five interwoven strands of mathematical proficiency: conceptual understanding, procedural fluency, strategic competence, adaptive reasoning, and productive disposition. The strands are interdependent—each a different aspect of, but all necessary for, mathematical proficiency.

Crucial to all of these strands is *metacognition*, or the ability to monitor our own mathematical thinking, including the degree of our understanding, our ability to connect mathematical ideas, and our capacity to think logically about mathematical patterns and relationships. These are "the glue that holds everything together, the lodestar that guides learning" (National Research Council 2000, 129). In other words, to be mathematically proficient, young learners need to develop the *thoughtful numeracy* described in Chapter One.

As educators teaching young mathematicians to develop these characteristics, we can use Guided Math conferences with students as a way of encouraging them to describe their understanding, justify their mathematical reasoning, and reflect more deeply on their mathematical comprehension (or lack thereof)—all indispensable parts of metacognition. Furthermore, during conferences, students gain confidence in their mathematical prowess as teachers provide constructive, specific feedback on their accomplishments and ask students to engage in the task of self-assessment. These intimate conversations about their work lead young mathematicians to see mathematics as something that often requires much thought and struggle, but that ultimately makes sense.

The Standards for Mathematical Practice

The Standards for Mathematical Practice recently set forth in the Common Core State Standards for Mathematics (2010) are based on both the strands of mathematical proficiency specified by the National Research Council and the process standards delineated by the National Council of Teachers of Mathematics (2000)—problem solving, reasoning and proof, communication, connections, and representation. They "describe ways in which developing student practitioners of the discipline of mathematics increasingly ought to engage with the subject matter as they grow in mathematical maturity and expertise throughout the elementary, middle and high school years" (Common Core State Standards Initiative 2010, 8).

Many teachers applaud the heightened focus on mathematical practices and are searching for ways to effectively instill and assess these habits of mind in their students. Traditional instruction has focused more on teaching and assessing content knowledge. According to Christinson (2012, 72–73), one of the implications of implementing the Standards for Mathematical Practice is a "shift in the classroom environment from a teacher-centered approach focused on teacher explanations to a student-engaging environment that utilizes student discourse, explanation, verification, collaboration, and metacognition to develop understanding of mathematical concepts." As a support for such a shift, students' one-on-one conversations with teachers about their personal mathematical understanding and work are invaluable.

With these student-teacher conferences, the evidence of student engagement—which is integral to the mathematical practice standards—is more visible. Teachers' feedback and questions lead students to become more adroit with the practices and processes of mathematicians. They practice the language of mathematical argument, conjecture, and discourse in risk-free conversations with teachers who know how to gently ask the questions that "poke holes" in faulty reasoning. They understand that they are accountable for the work they do—that they are expected to do more than simply give an answer to a question. As participants in math conferences, they know that their teacher cares about their mathematical thinking and demands that they think critically about their learning as preparation for life.

Knowing What Matters: Learning through the Eyes of Young Mathematicians

As with any worthwhile endeavor, it is wise to step back and reflect at times. We grow as educators when we reflect on exactly what we know about the world of mathematics as our students experience it, how we come to know about our students' experiences and perceptions, and how we can learn even more about the ways in which they think about and use math, both in school and in the world around them.

Traditionally, limited insight into students' mathematical abilities has been provided to teachers by written tests, homework, activity sheets, and class participation. For their part, students have typically depended upon their teachers to inform them as to how well they are learning math, assuming very little responsibility for monitoring their own learning. As educators, we have placed little responsibility on students for reflecting on and monitoring their progress.

Recent research, however, indicates that learning more about what our students know and do not know, what they can and cannot do, along with what they are thinking as they work, gives us the building blocks we need as teachers to maximize student learning (Andrade 2010; Black and Wiliam 2010; Fisher and Frey 2007; Hattie 2012b; Stiggins 2005, 2007). Not only that, but by involving learners themselves in the process of assessing their understanding and skills, they become better learners. It is really a matter of knowing what matters.

Guided Math Conferences for Formative Assessment

It seems obvious that the more we know about our students' learning, the more accurately we can tailor instruction to meet their needs. As the National Council of Teachers of Mathematics (2000, 22) put it, "Assessment should not merely be done to students; rather, it should also be done for students, to guide and enhance their learning." Unfortunately, the pervasiveness of standardized, high-stakes testing has caused teachers and students to at times view assessment with a degree of trepidation. Rather than emphasizing assessment *for* learning, the nature of these tests shifted public focus to assessment *of* learning as a measure of accountability for students, educators, and schools.

With the quest to prepare students for "the big test," a powerful means of assessing mathematical understanding and competence is often overlooked. Personal communication between teachers and students is one of the most productive means of assessment. "Our personal exchanges with students can be packed with useful information about their achievement, and thus can serve a variety of important and often interwoven purposes" (Stiggins 1997, 225). Guided Math conferences are opportunities for teachers and students to "talk directly and openly about levels of student attainment, comfort with the material to be mastered, specific needs, interests, and desires, and/or any other achievement-related topics that contribute to an effective teaching and learning environment" (241). Additionally, with these one-on-one conversations, assessment and instruction can be one and the same. The assessment experience itself can be a powerful instructional tool for teachers and students.

While summative assessments—assessments *of* learning—certainly have an important role in education, more important are formative assessments—assessments *for* learning. These inform our instruction, making our teaching more efficient and effective. Fisher and Frey (2007, 4) define formative assessments as "ongoing assessments, reviews, and observations in a classroom" that are used to "improve instructional methods and provide student feedback throughout the teaching and learning process."

Research on the use of formative assessment shows that student gains resulting from the use of this type of assessment are among the largest reported for any educational intervention. Moreover, these gains were particularly pronounced for low-achieving students. Because of the results of this research, Black and Wiliam (2010) conclude that efforts to strengthen the practice of formative assessment produce significant, and many times substantial, learning gains for students.

Guided Math conferences provide just the kind of information teachers need to decide how to proceed with instruction, and with opportunities to combine assessment with instruction targeted to students' immediate learning needs.

What Is Formative Assessment?

What constitutes a formative assessment? Chappuis (2009, 6) enumerates the following conditions which an assessment must meet in order to be considered formative rather than summative in nature:

- It aligns with the standards to be learned.

- Its criteria matches what has been or will be taught.

- It provides sufficient detail to pinpoint needs and guide future instruction.

- Its results are timely.

- Teachers and students act in response to the results.

With the use of Guided Math conferences for assessment, these conditions are easily met. Students are working on mathematical tasks selected to support the learning standards being taught and are thus aligned with the standards. Conversations generated between the teacher and student focus on assessing the student's degree of understanding, skills aligned to the standard being taught, and/or the student's disposition toward mathematics.

Hattie (2009) persuasively stresses the importance of teachers evaluating the effectiveness of their teaching and then, as a response, altering the direction of learning to meet challenging, shared teacher-learner goals. Hattie (2012b, 15) states, "Learners can be so different, making it difficult for a teacher to achieve such teaching acts: students can be in different learning places at various times, using a multiplicity of unique learning strategies, meeting different and appropriately challenging goals." It is precisely because of this difficulty that Guided Math conferences are of such value to educators. Guided Math conferences allow teachers to discover how effective their teaching has been through conversations with individual students and then to teach in response to their findings. Because of this, conferences are an important instructional and assessment tool for teachers striving to meet the "six signposts towards excellence in education" Hattie (2012b, 18) identifies in his synthesis of more than 900 mega-analyses of research. One of these signposts states:

> *Teachers need to be aware of what each and every student in their class is thinking and what they know, be able to construct meaning and meaningful experiences in light of this knowledge of the students, and have proficient knowledge and understanding of their subject content so that they can provide meaningful and appropriate feedback such that each student moves progressively through the curriculum levels.*

Conferring with students about mathematics is particularly effective in providing *timely* information about *specific* strengths and needs that lead both teachers and students to take the next steps to address identified needs, whether they are correcting misunderstandings, filling foundational gaps in knowledge or skills, or extending understanding with additional challenge.

According to Stiggins (2007, 24), formative assessment of the kind possible with conferring "provides both students and teachers with understandable information in a form they can use immediately to improve performance." More than just a "one-time event stuck at the end of an instructional unit," it is instead a part of "interlaced experiences that enhance the learning process by keeping students confident and focused on their progress."

Guided Math Conferences and Student Self-Assessment

Over time, theories about best practices in education have shifted away from those of simple transmission of knowledge from teacher to student. We now believe it involves a process through which students construct their own knowledge. Through interactions with subject matter and focused discussions with others, learners connect their experiences to their prior knowledge and make meaning (Nicol and Macfarlane-Dick 2006). Despite these changes in what we understand about teaching and learning, teachers continue to have the primary responsibility for assessing students and providing them with feedback.

The Value of Self-Assessment by Young Mathematicians

Where am I going? How am I going? and Where to next? An ideal learning environment or experience occurs when both teachers and students seek answers to each of these questions (Hattie and Timperley 2007, 88).

The quote above emphasizes the important roles of not only teachers, but students, too, in seeking answers to these three questions about learning. Each of these questions relates more personally to students than to their teachers. Thus, student learning is more meaningful and enhanced when learners and teachers are partners in the process of assessing *for* learning.

Research from a wide range of content areas supports the notion that the responsibility for formative assessment should be shared not only by teachers, but by students as well (Andrade 2010; Black and Wiliam 2010; Butler and Winne 1995; Carr and Biddlecomb 1998; Schunk 2003). The active role of students in assessing their own work and monitoring their progress toward well-established learning goals has been linked to significant improvements in performance for all students (Andrade 2010), particularly those who struggle the most (Black and Wiliam 2010). Through the reflective process of self-assessment, students develop a better understanding of the purposes of their learning and become more focused on what they need to do to be successful.

Nicol and Macfarlane-Dick (2006, 200) conclude, "If formative assessment is exclusively in the hands of teachers, then it is difficult to see how students can become empowered and develop the self-regulation skills needed to prepare them for learning outside [school] and throughout life." Self-regulated learning as referenced above is defined by Andrade (2010, 5) as "a dynamic process of striving to meet learning goals by generating, monitoring, and modifying one's thoughts, feelings, actions and, to some degree, context." Indeed, as such, self-assessment is a central component of an individual's ability to self-regulate. The acts of generating, monitoring, and modifying thoughts and actions to meet learning goals are key to self-assessment. So not only is formative self-assessment an important academic skill for students, it is also an essential lifelong-learning skill.

Features of Self-Assessment

Self-assessment might best be described as a process of formative assessment during which students assume the responsibility of reflecting on the quality of their work as compared to established criteria, and monitor the degree of their understanding of both content and procedural knowledge. According to Andrade (2010), this process is comprised of three steps.

1. Teachers, students, or both, articulate expectations for learning or for the work product.

2. Students critique their own understanding or work in terms of these expectations.

3. Learners use the feedback they generate from the self-assessment to revise their work or take steps to increase comprehension.

This final step is what moves the self-assessment process into the realm of formative assessment. In essence, "students become both self-assessors and consumers of assessment information" (Stiggins 2007, 24). Without the chance to act on what they learn from the process, the assessment is not formative in nature.

Gregory, Cameron, and Davies (2011) clearly distinguish self-assessment from self-evaluation. Self-evaluation involves gathering information about learning and making judgments about it—assigning numbers, percentages, or grades. In contrast, when students self-assess, they describe their work and their thinking, compare it to established criteria for quality work, rate their understanding of the content, and decide where to go from there. Rather than focusing on percentages or grades, students "see their strengths, understand what they need to work on, and are able to set personal goals" (60).

While the focus of much of the literature on self-assessment centers on assessing student work products, of equal importance is the role of students in identifying their own misconceptions and monitoring their conceptual understanding. Eventually, a lack of understanding may affect the quality of the work product and become obvious at that time, but more timely recognition of a breakdown in comprehension is imperative. Erroneous practices employed by students working under misconceptions may be difficult to unlearn. When students develop the capacity to carefully monitor their comprehension and then take steps to remediate their understanding when needed, learning is enhanced.

Fostering Self-Assessment

If learners are to actively examine and self-regulate their own learning trajectory, they need guidance. Most have never before been expected to assume this responsibility. To facilitate this, we must "give students supports that they can hold on to as they take the lead—not just push them onto the path and hope they find their way" (Fisher and Frey 2008, 32).

First, it is vital that students understand the value of the self-assessment process itself. How will it lead them to greater achievement? Students should know the purpose behind their work, the learning expectations, the criteria for success, how their work relates to the standards being studied, and finally, strategies for repairing breakdowns in comprehension. Moreover, students should have personal learning goals, so that they are able to check on their progress toward a specific target. To view their work and mathematical comprehension critically, students must also possess the relevant vocabulary of assessment and the habits of thought that lead to effective self-assessment.

Modeling is an especially useful method for supporting learners as they embark on the self-assessment path. Guided Math conferences are ideal venues for students' initial efforts at self-assessment. Teachers need to model the self-assessment process, working closely with students as they examine work samples and justify their thinking. Teachers can model their thinking as they consider how a work product measures up against defined criteria and learning goals. Or a teacher can pose questions to help reveal students' mathematical understanding or lack thereof. Through these experiences, students are learning how to independently assess their own work and understanding. By giving students descriptive feedback, teachers model for them the kind of thinking they should engage in when looking at their own work. When teachers ask students what actions they will take to address the feedback, they are teaching them how to take control of their learning. Once students acquire the foundational knowledge and skills needed for reflecting critically on their work, they should be expected to share their own take on their learning, provide evidence to justify that take, and then share their plan to extend their learning.

In one-on-one meetings, teachers can nudge students gently yet firmly to assume greater responsibility for assessing their achievement, providing justification for their assessment, giving themselves feedback, and setting their own learning goals. With practice, students should be able to explain where they are going, how they are going, and where they will be going next. When students realize how closely self-assessment and goal setting are linked to the learning process, most take it seriously and become skilled at monitoring their work (Gregory, Cameron, and Davies 2011). The more adept learners become with the process, the more "they develop an internal sense of control over the conditions of their success and greater ownership of the responsibility for improving" (Chappuis 2009, 95). In essence, they become self-regulated learners.

Five Strategies for Effective Guided Math Conferences for Assessment

To make Guided Math conferences effective as a formative assessment tool for both teachers and students, teachers should consider these strategies:

1. **Help students develop a clear understanding of their learning goals and how they will know when they meet those goals.**

   Before teachers can convey learning expectations to students, they must have a clear understanding of what they are in their own minds. What is required for a student to master a given standard? What is the evidence of success? Can the overarching learning goal be broken down into smaller, more manageable chunks? If so, what exactly should be the immediate target for the student with whom they are conferring? What if he or she achieves that goal? What is the next learning goal?

   When teachers are clear in their own thinking about the desired learning goals and evidence of learning, they are able to reinforce them with students during these one-on-one discussions about the work in which they are presently engaged and lead them to set goals for themselves.

2. **Guide the conversation with questions to elicit evidence of student learning, both content and process, and/or misconceptions and gaps in foundational knowledge and skills.**

   Open-ended questions prompt students to explain their work or justify their reasoning and encourage them to reflect on their own thinking processes. Even those of us who are experienced teachers have to guard against assuming we know how our students think. Sometimes, we examine a student's work and from our past experiences make snap judgments about what the student was thinking. We may assume that arriving at a correct solution from a very capable student is evidence of understanding. Or we may see a common error and assume the student holds a common misconception that often leads to that error. While these assumptions may be true in some cases, good questions that reveal the student's mathematical reasoning may expose hidden misconceptions; alternative, but valid, ways of approaching the mathematical work employed by the student; or simple computational mistakes when otherwise the mathematical reasoning of the student is fully justified.

3. Encourage students to reflect on their mathematical understanding so that they assume ownership of their learning.

Research shows that when students actively engage in monitoring and regulating their own learning, the rate of learning increases dramatically (Carr and Biddlecomb 1998). These students monitor their thinking to determine whether they have met the established criteria for mastery. In effect, they begin to give themselves descriptive feedback and apply strategies to repair their mathematical comprehension, when necessary. Allowing students time to pause and think during conferences makes it more likely that they will thoughtfully reflect on their own learning and check their progress toward their learning goals (Gregory, Cameron, and Davies 2011).

4. Provide specific feedback to let students know both what they are doing well and what will move them forward in their mathematical learning.

It is both motivating and affirming for students when they hear exactly what they have done well, in which ways their mathematical thinking is valid, or in which ways they have grown mathematically, as opposed to no feedback or to more generic compliments like "good work" or "you have the correct answer." The positive feedback students receive may even surprise students and make them aware of aspects of their work that they did not realize was done well. Positive recognition encourages young mathematicians to continue to do whatever they have done well and to work to garner more compliments on their work by doing the best they can. It also creates a nurturing learning environment in which students are ready to consider feedback that will lead them to their next steps in learning. The "next step" feedback is what promotes student growth as mathematicians. Because it occurs as they work, students are able to immediately reconsider their approach and/or extend their thinking.

5. **Use the information gathered during the conference to identify a teaching point to move student learning forward.**

What ultimately defines formative assessment is the way in which we use it to drive our instruction. Formative assessment and teaching are complementary processes, one supporting the other (Sammons 2010). The specificity of information attained through this type of assessment is crucial in steering us as we make day-to-day teaching choices. Information about learning needs identified during conferences may be used immediately as a teaching point during the conference or later when planning small-group lessons or even large-group experiences. Because of the value of this assessment data, it is important to develop and maintain a readily accessible system of anecdotal notes to document what has been discussed during conferences. Notes should also include what was taught in response to observed learning next steps.

Empowering Young Mathematicians with Effective Feedback

As alluded to in the previous section, feedback plays a major role in formative assessments—whether these assessments are conducted by teachers or are student self-assessments. Effective feedback may lead to increased achievement by providing recognition of quality work, to intervention if errors or misconceptions are present, or to teacher guidance that points students toward their next steps in learning. At its best, feedback motivates students to manage their progress toward clearly articulated learning goals and places the emphasis on learning rather than on simply attaining a good grade.

What Is Feedback?

Most of us are familiar with the irritating and often screeching feedback noise produced when a microphone is too close to a speaker. The jarring output is part of a positive feedback loop in which the sound from the speaker is amplified by the microphone to increase the sound of the speaker, which in turn is amplified again, looping over and over. This is *not* what constitutes positive or effective feedback for learners!

According to Wiliam (2012), feedback for students is similar to the negative feedback loop of a room thermostat. With this feedback loop, an increase or decrease in temperature triggers a response in the heating system. The temperature drops below a certain point, and the heat is turned on. Once the temperature reaches a certain point, the heat turns off. This feedback is part of a *system* that regulates the temperature of a room. Without the feedback, the system might either run nonstop, would be off at all times, or would periodically turn on and off with no regard to the temperature of the room—all of which might well result in temperatures that are uncomfortably warm or cool. Without a reaction to the feedback, the same consequences ensue. Together, the feedback and the reaction to the feedback produce the desired goal—a comfortable room. In providing feedback to students, we also hope to help our students reach a desired learning goal. Our feedback alone will not make the difference; the response it elicits from learners is what gives feedback its powerful impact.

At its most basic, feedback is just "information about how we are doing in our efforts to reach a goal" (Wiggins 2012, 11). Just as coaches critique the performance of athletes during practice, offering observations and tips for improvement, so teachers can help students identify where they are on their personal learning trajectory and prompt or cue them to take the next steps in learning.

The evidence from decades of research shows that feedback can have a significant positive effect on student learning (Hattie 2009; Marzano 2007). A number of studies, however, have found that some feedback, particularly in the form of scores or grades, can reduce student learning or even worsen learners' performance (Butler 1988; Dweck 2000, 2006; Kluger and DeNisi 1996). What appears to make the difference between effective feedback and ineffective or even harmful feedback is the kind of responses triggered in individual students rather than the nature of the feedback itself (Wiliam 2012).

While there is no simple prescription for offering effective feedback for all students, the personal nature of Guided Math conferences during which students share their mathematical thinking with their teacher makes this a valuable setting for feedback. Conferring individually with students, teachers are able to not only determine student strengths and needs, but also deliver specific feedback in a way that motivates students to respond in productive ways that enhance both their mathematical work and understanding. The prompt feedback students receive gives students the information they need to adjust their learning approach and take ownership of their learning (Van de Walle, Karp, and Bay-Williams 2010). Moreover, students may be encouraged to assess their own work and generate their own feedback. Ongoing conversational links during conferences allow teachers to ensure that students "digest, understand, and use the feedback" (Brookhart 2012, 25).

Characteristics of Effective Feedback

Because effective feedback can lead to substantial gains in student learning, while ineffective feedback can negatively impact learning, it is important to know which aspects of feedback are associated with gains in student learning. Most researchers and educators agree on these characteristics of effective feedback:

- **Effective feedback is clear and specific, focusing on things that are within the learners' control.** Ideally, the feedback is specific enough to guide students to their next steps, but still leave the thinking to them (Chappuis 2012; Wiliam 2012). It should be so tangible and transparent that students can readily learn from it—helping them recognize areas of growth or need that may not be apparent to them (Wiggins 2012).

- **Effective feedback is user-friendly, couched in such a way that students understand it.** No matter how specific feedback may be, its value is limited unless learners understand it (Wiggins 2012). Much teacher-supplied feedback is never fully received. According to Hattie (2012a, 20), "Students typically miss the teacher's messages, don't understand them, or can't recall the salient point."

- **Effective feedback is honest and respectful.** Providing effective feedback requires actively listening as students describe their work and thinking. Then, it faithfully describes strengths and needs while, at the same time, conveying to students their teachers' regard for them as active learners and fellow mathematicians on a learning quest (Brookhart 2012).

- **Effective feedback is delivered in a timely manner, but may not always be immediate.** Feedback is timely in the sense that students receive it when the work is still fresh in their minds and when there is time to act on it (Brookhart 2012; Chappuis 2009, 2012; Wiggins 2012). Yet learners sometimes need the chance to struggle and make errors—and eventually come up with valid solutions to their struggles. Immediate feedback may shelter them from such valuable learning experiences. Effective teachers use their professional judgment in deciding the "just right" time to confer and provide feedback—allowing time for experiential learning, but preventing errors and misconceptions from becoming practiced and engrained.

- **Effective feedback prompts student thinking rather than doing it for them.** Students have no need to think for themselves when teachers provide more guidance than is needed; that kind of feedback fails to deepen student learning (Chappuis 2012). Effective feedback, however, requires more work on the part of students than on the part of teachers (Wiliam 2012). Feedback can powerfully guide student thinking through the use of carefully crafted questions directing students to discover their own errors and misconceptions or extend their thinking.

- **Effective feedback is differentiated in response to student learning needs.** Not only should feedback be differentiated based on identified academic strengths and needs, but also based on teachers' knowledge of students' learning styles and personalities. Some students may only need a reminder for the next step, while others may need more explicit prompts or examples (Brookhart 2012).

- **Effective feedback addresses partial understanding by students.** According to Hattie (2012b), feedback does not occur in a vacuum; it follows instruction. Without at least partial understanding, feedback may well have a detrimental effect (Chappuis 2009, 2012; Hattie and Timperley 2007). Feedback builds on prior knowledge and skills. If students lack a minimal level of either knowledge or skill, they will most likely fail to understand the feedback. In these cases, it may be necessary to provide further instruction instead of giving feedback. This may be in the form of a teaching point during a conference, a small-group lesson, or when knowledge is lacking for most of the class, a whole-class lesson.

- **Effective feedback directly references and aligns with the intended learning target by describing student work or thought processes in relationship to those learning goals.** The feedback itself may help clarify a goal in order to reduce the gap between where the student is and where he or she should be. At times, teachers may need to model what success looks like with worked examples or demonstrations. Students should be encouraged to assess their own work and understanding based on both class and individual learning goals.

- **Effective feedback addresses quality of the performance, the work product, or conceptual understanding rather than the students themselves.** As Wiliam (2012, 34) puts it, the most effective feedback "focuses on the task at hand rather than the recipient's ego." Imperative in providing feedback that works is considering how the recipients react to it. When the feedback describes the work or understanding—not the student personally—it supports a risk-free learning environment in which students measure their success by the progress they make rather than in comparison to others. While being compared to others too often triggers a concern for well-being at the expense of growth (Wiliam 2012), task-specific feedback points young mathematicians in the right direction and encourages them to take the next step in learning.

- **Effective feedback is actionable and focused on one or at most two things students can use immediately to improve their performance or understanding.** Generic evaluative statements such as "Nice job!" or "You can do better" offer students no ideas for next steps in learning—neither how to extend and enhance learning based on what they did well nor how to improve their work or knowledge. Instead, "effective feedback is concrete, specific, and useful" (Wiggins 2012, 14). On the other hand, too much corrective feedback at one time may cause students to be overwhelmed and unable to act on any of it; they simply shut down. Chappuis (2012, 39) recommends that teachers let "go of the urge to make the work perfect and instead provide as much guidance as the student can reasonably act on." Teachers find that feedback is most useful to learners when it focuses on a single aspect of a student's product, performance, or conceptual understanding—one that is most likely to make a significant difference in achievement. Therefore, it most often addresses errors rather than simple mistakes. Although mistakes are largely a consequence of inattention or fatigue, errors occur because of lack of knowledge (Fisher and Frey 2012). Directing attention to the source of student errors highlights precisely what they need to do to improve their learning. This approach prevents students from being overwhelmed or distracted by multiple learning points.

Establishing Student Learning Goals

Before teachers can assess student progress and offer feedback in Guided Math conferences and before students can self-assess their progress, learning goals must be clear. Learning goals are distinct from learning activities. A learning goal states what a student will know or be able to do following instruction, whereas learning activities or assignments are the actual tasks a student does during instruction (Marzano 2007). For example, the statement *Students will create a three-step word problem requiring at least two different operations* is a learning activity. The statement, *Students will understand and be able to apply the distributive property in computation* is a learning goal. One-on-one conferences between teachers and students are ideal for teachers to both reinforce goals previously shared with students during whole-class or small-group lessons and to lead students to develop their own more personalized learning goals.

Hattie's (2012b, 27) research suggests that in order for teachers to substantially influence the learning outcomes of their students, "challenging goals, rather than 'do your best' goals" must be established, and their students must be committed to achieving these targets. Such clearly articulated goals are an essential component of both student motivation and learning.

Research shows that learners who are committed to attaining a goal are more likely to measure their performance against the goal as they work (Schunk 2003), thus providing themselves with valuable feedback. As a result, this self-assessment raises their self-efficacy and sustains motivation as they work toward proficiency. These learners also tend to plan how they can achieve their goals as they monitor their learning progress and as a consequence do better in school than students who do not have established goals (Andrade 2010). Undoubtedly, goal setting enhances student achievement, which may be why Tomlinson (2013, 89) includes helping students of poverty learn to set learning goals and take action to meet those goals as one of the central practices of teachers "who stare down poverty."

Research also suggests that the types of goals students have influence their learning success. Students whose goals related specifically to the academic "conditions, products, and standards of their studying learned more than students who either set goals related to the amount of time spent studying or general goals such as 'learn the material'" (Andrade 2010, 7).

Guided Math conference conversations allow teachers to have purposeful conversations in order to informally clarify learning goals for mathematics, so that their students clearly understand exactly what these goals entail. Furthermore, during these face-to-face discussions, teachers can not only encourage students to set their own individual learning goals to address unique needs and learning styles but they can also influence the kinds of goals learners set—making them more content-specific, measurable, and achievable.

The assessment and goal setting process is, in fact, cyclical. Although learning goals are necessary for assessment—how well have the established goals been achieved? —goal setting is also the logical next step following assessment. Taking into account the results of the assessment (whether by teacher or self), students reflect on their current needs. They may opt for short-term goals, which are small and specific, or for long-term goals based on a broader area in which they wish to improve, or both (Gregory, Cameron, and Davies 2011). An important benefit of conferring with students is the opportunity for teachers to engage students in honest discussions about their learning, and then to nudge them along as they set their own personal learning goals. When else during busy school days do teachers have the luxury of these informal but highly focused chats with students regarding—what is, after all, the ultimate aim of teaching—learning?

Promoting Accountability

Another important aspect of math conferences is holding students accountable for their work. More than a simple answer to a problem is expected when teachers and students confer. Young mathematicians are asked to give evidence of their learning. The focus shifts from number of correct or incorrect answers on a test or worksheet to a collaborative examination of broader indications of understanding and skill. Bestowing this responsibility on students helps them develop a sense of ownership in their own learning (Davies et al. 1992).

Students may demonstrate their understanding by explaining their problem-solving strategies, telling what they know about a math concept, or explaining their reflections from their math journals. According to Stiggins (1997, 18), students who engage in a thoughtful analysis of their work in which they "identify its critical elements or internalize valued achievement targets become better performers."

Students may initially fear the risk associated with being held accountable for their learning and performance, preferring to offer minimal insights into their thinking and being reluctant to reveal confusion or lack of understanding. After all, they rightfully recognize the fact that they are being assessed, however informally. Assessments have a powerful effect on their academic and personal self-concepts (Stiggins 1997). However, we need to keep in mind that students are more likely to feel in control over their well-being when they know what the criteria for success is and know exactly what the learning target is. When we create a risk-free environment for students in which they are held accountable yet are comfortable acknowledging confusion or uncertainty, we give them the power, along with the responsibility, to monitor and maximize their learning.

Teaching and Learning

Much has been said earlier in this chapter about assessment, feedback, and goal-setting—all of which are powerful tools for promoting student learning. Guided Math conferences offer teachers ample opportunities to employ these tools—and also for providing instruction based on information gleaned through their use. In fact, since teaching is often at its most effective when it is directly linked to these three tools, the instruction students receive during conferences can sometimes accomplish more than a longer small-group or whole-class lesson.

In the characteristics of effective feedback described earlier in this chapter, it was noted that effective feedback addresses partial understanding. Without at least partial understanding, reteaching is necessary. This may be addressed with small-group or even whole-class lessons when a number of students need further instruction. Many times, however, the student-specific reteaching can be done during a Guided Math conference.

Since one-on-one interactions with students are such productive sources of insight into the thinking of students, teachers are able to identify learning "next steps" as teaching points for conferences and then deliver brief, targeted lessons immediately. As teachers, our role during a Guided Math conference is to discover the mathematical thinking of our students, provide feedback, and then correct, clarify, and/or extend that thinking in a way that empowers our students as learners. By asking ourselves, *What is the one thing I can teach this student right now that will help the most?* we can provide timely and differentiated "just right" instruction for students. This limits the amount of corrective information we offer to an amount the student can successfully act on (Chappuis 2012).

What does research show about the efficacy of instruction during face-to-face conversations as compared to teaching an entire class of students? We tend to assume that instruction is more efficient when delivered to a group of students—making the additional assumption that if lessons are effective in one-on-one conversations with students, they should also work with a group of students (Nuthall 2005). This belief fails to take into account, however, the nonverbal and verbal feedback (nods, expressions of confusion, questions, eye contact) we receive constantly from conversational partners when we are involved in one-on-one conversations. These signals let us know how well our message is being understood. In response to these subtle—or sometimes not-so-subtle—messages, we adapt our lesson to head off misunderstandings or misconceptions and "repair" our conversations (Hutchby and Wooffitt 2008). Although spontaneous interactive repair is doable during one-on-one conversations, as the number of participants in the discussion increases, the difficulty in responding to multiple, and sometimes conflicting, messages being given by students increases. As a result, the degree of student learning tends to decrease.

Teachers attest to the benefits of conferring with their students about mathematics. One example is Kira Walker, a teacher at Brockton Elementary School in Los Angeles who engaged in a two-year long inquiry project on teacher/student math conferences. She reflected that conferences "informed my teaching, and I have a much deeper understanding of where communication may break down. In addition, the student/teacher conferences allowed me to assess my students in a way that standard whole group instruction never could" (Teacher Learning Collaborative, final paragraph).

Chapter Three will provide additional information on the structure of math conferences and how to deliver brief lessons based on student needs.

Encouraging Mathematical Communication

Over a decade ago, the National Council of Teachers of Mathematics (2000) identified reasoning and proof, and communication as two of the process standards in their Principles and Standards for School Mathematics. They proposed that students should learn that math makes sense and that mathematical assertions should always have reason to back them up. They need to develop the ability to reason and to justify that reasoning, both of which are essential to understanding the discipline. NCTM (2000, 61) also advised educators that "[s]tudents gain insights into their thinking when they present their methods for solving problems, when they justify their reasoning to a classmate or teacher, or when they formulate a question about something that is puzzling to them."

These skills are difficult to promote in a classroom full of young learners. Whole-class discussions rarely offer more than a few students the chance to express their ideas. Having students share with a buddy is an option, but in those scenarios they get little teacher feedback, and sometimes the feedback given by the buddy may be erroneous. The one-on-one nature of Guided Math conferences, however, offers a format in which teachers can encourage students to engage in this type of discourse with them.

When conferring with young mathematicians, teachers can probe for explanations of strategies and processes. Students learn that they are expected to communicate about their mathematical thinking and gain verbal fluency through this guided practice with their teachers. In their part of the dialogue, teachers model the use of mathematics-specific vocabulary and how to clearly and precisely express mathematical ideas. Participating in these experiences, students learn what constitutes "math talk" and how well-reasoned justifications are composed.

More recently, the Standards for Mathematical Practice, an integral part of the Common Core State Standards for Mathematics (CCSSI 2010), also places an emphasis on the instructional value of interactions between teachers, students, and the mathematics. Teaching these standards requires teachers to create classroom environments "that utilize student discourse, explanation, verification, collaboration, and metacognition to develop understanding of mathematical concepts" (Christinson 2012, 73). While teachers may implement these standards in many different ways, math conferences provide a viable structure for encouraging students to delve deeper into their mathematical thinking and to express their reasoning in a fluent and coherent manner.

Building Relationships

"Arguably the quality of the relationships teachers have with their students is the keystone of effective management and perhaps even the entirety of teaching" (Marzano 2007, 149). This is a very strong statement, but it is one that most teachers intuitively believe to be true. In his meta-analysis of data, Marzano found that there are two complementary aspects of effective teacher-student relationships—the sense the teacher gives of providing guidance and control, and the sense the teacher projects that he or she and the students are a team working for the well-being of the entire class. In other words, the teacher communicates that he or she has a personal stake in the success of each student. And when the teacher-student relationship is good, then everything else within the learning environment seems to be enhanced.

As teachers, how can we convey to our students the sense of community and concern for the success of all? According to Marzano (2007), teachers' behavior is their language of relationship—students interpret their teachers' behaviors as statements about the kind of relationship their teachers want to have with them.

This is especially telling when we consider the typical way in which teachers talk with students. When Hoffer (2012) spent a day listening to and then tallying all the teacher-student dialogue she heard, nearly all the conversations fit into these predictable categories, in order from most to least common:

- Policing

- Directing

- Accounting

- Correcting

- Rescuing (i.e., go back and fix your answer)

- Socializing

- Praising

The authoritative nature of these interactions profoundly affects students' relationships with their teachers. The conversations teachers share with students during conferences, on the other hand, help to build more positive relationships.

Strong relationships with their teachers are especially important for students from low-income families. Contrary to common belief, research shows that for these students the primary factors in their motivation and achievement are the school and the teacher rather than the home environment (Jensen 2013; Irvin et al. 2011). In light of this, Jensen advises teachers to find ways of establishing strong relationships with their students from impoverished homes by revealing more about themselves and learning more about their students (Jensen 2013). One-on-one conference conversations provide teachers time to do just that. "To confer is to intentionally create opportunities to honor learners' ideas by listening and learning alongside them.…Conferring is a key forum for conveying to learners our deep respect for their learning" (Hoffer 2012, 139).

When we choose to take time to confer with our students about their mathematical work and understanding, we send a clear message that we care. When we listen intently and share our thinking with them as one mathematician to another, we display respect and concern for their success. Carl Anderson (2000, 23) describes the importance of writing conferences:

> *A few words, a smile, a nod of understanding. That's all it takes to show students we care about them. That's all it takes to inspire some students to stretch themselves as writers. That's all it takes to change some students' writing lives.*

Although Anderson was referring to the power of writing conferences, it is equally true for Guided Math conferences. Our personal interest in our students and their understanding inspires learners to take risks and stretch themselves as mathematicians. We demonstrate our concern for them and their success with mathematics when we make time to confer with them, converse with them, and support their fledgling efforts to understand a complex and challenging discipline.

Chapter Summary

As demands for greater depth and rigor in mathematics education increase, the rich interactions between teachers and students as they confer are conducive to thoughtful reflection, intellectual curiosity, and motivation for learning.

Guided Math conferences are an effective instructional vehicle for teachers to apply many of the practices that research shows are instrumental in promoting learning success. Studies highlight the positive impact of formative assessment by teachers and students, timely descriptive feedback, and goal setting—all of which are integral ingredients of math conferences.

The feedback that many students receive on their mathematics work consists of a grade on a worksheet or test. They learn how many of their responses were correct and incorrect. These learners see the grade as the measure of their success, but have little motivation to improve their performance or feedback on how to improve, even if they wish to.

Legendary football coach Vince Lombardi once said, "Practice doesn't make perfect. Only perfect practice makes perfect." Yet in classrooms today, too many of our students are working diligently, but practicing mathematics using faulty strategies and incorrect procedures based on misconceptions. To ensure success, athletic coaches intervene during practice to correct errors, rather than waiting until after the game. Chappuis (2012, 38) queries teachers, "[H]ow long do we allow students to repeat a mistake or cement a misconception?" Feedback is most effective *during* the learning cycle, not *after* it. Teachers who confer with their students on a regular basis are more likely to catch the mathematical errors of young mathematicians before they become habitual.

Impromptu lessons based on learning "next steps" discovered by teachers and students collaborating during math conferences are timely and targeted. Students know they are accountable for justifying their mathematical work and for clearly expressing their thinking. Teachers and students are able to share their questions and their reasoning—mathematician to mathematician. The bonds between teacher and student are strengthened based on the teacher's demonstration of care and respect.

REVIEW AND REFLECT

1. How have the demands for increased rigor and depth in mathematics education affected your teaching? In what ways can you use math conferences to meet these demands?

2. How does the use of math conferences for formative assessment compare to your current methods of formative assessment?

3. When you identify the "next steps" in learning for individual students in your class, how do you provide the needed instruction? What are the advantages/disadvantages of that method of instruction? How does the use of math conferences compare to your current method?

The Structure of a Guided Math Conference

Perhaps some of the most challenging aspects of effective mathematics instruction, and particularly of implementing the Guided Math framework, are the tasks of assessing student understanding and skill and then planning how to guide individual students to their appropriate next steps in learning. Paper-and-pencil tasks may provide some indications of exactly what students know and can do, but at times, students' answers may be correct for the wrong reason or because they glanced at the work of a neighbor. Incorrect responses may be due to simple computational errors or may reflect a lack of understanding. These student errors tell us little about whether there are actual breakdowns in understanding, or if there are breakdowns, precisely where they occur. Without specific information about the causes behind the erroneous responses, how can teachers prepare to target students' learning needs accurately?

Traditionally, observations from class discussions have supplemented the assessment data from quizzes, tests, and projects. The structure of these conversations, typically conducted with the whole class, most often follows a pattern of question-response-evaluation, even for more challenging questions that go beyond the simple recall of information. This pattern of interaction reinforces the notion of the teacher as the dispenser of knowledge and the arbiter of whether responses are correct (Calkins, Hartman, and White 2005). As a consequence of this kind of discussion, students feel little responsibility for learning or for monitoring their own understanding—creating a very passive learning environment. And unfortunately, teachers learn little about the mathematical thinking behind the responses of their students. Lacking this important insight into student cognition, we find it challenging to decide how to differentiate instruction so that we address the unique learning needs of our students.

The pattern of conversation characteristic of Guided Math conferences contrasts greatly with the traditional pattern of class discussions. The conference is math talk between two mathematicians sharing their knowledge, ideas, and questions. We are able to probe the thinking of our students to discover the extent of their understanding and to uncover any misconceptions they may have. In this way, we become partners with our students on the learning journey. Students begin to see us, their teachers, as mentors with whom they feel comfortable sharing any confusion they may have about mathematical concepts or procedures. We can use carefully crafted questions to encourage our students to extend their understanding and to nurture their curiosity about the discipline of mathematics.

As these one-on-one discussions continue, students receive feedback on what they have done well and on ways to improve their work (Sammons 2010). The individualized nature of the teaching done in this setting allows the instruction to be quite specific to student needs. In addition, because of the close interaction between the teacher and student during math conferences, we can more easily assess how well students understand the teaching point we are conveying during the conference. And finally, we can emphasize the importance of our students continuing to make use of their mathematical understanding as we ask our students to restate and reflect on what they have learned.

Selecting Students to Engage in Guided Math Conferences

It is not always easy to know with whom to confer first. Who needs our support the most? Who is ready for a greater challenge? Who is working diligently, but under misconceptions about some aspects of mathematics? As we glance around our classrooms, we see students working, students distracted, students exploring. With whom can we, as teachers, have the greatest immediate impact? Which students need to confer with us the most?

In some instances, we may have in mind students with whom we feel we need to confer right away and others who can wait a bit. We may need to conduct follow-up conferences based on prior conversations or observations in conferences, small-group instruction, or large-group instruction. We know what we are looking for. We are checking to be sure that these students are applying strategies we have previously discussed. Or that they truly understand the concepts with which we are working. Or that these students are indeed digging deeper—increasing their mathematical prowess.

If the order of students with whom to confer is not clear, glancing around the classroom or listening to student chatter as they work may give us greater insight into which of them will benefit most from a math conference first. These observations let us know not only which students may be struggling, but also which students may need additional challenge (Sammons 2010). Often those students who very quickly demonstrate mastery of new concepts or strategies will need additional challenges. They benefit from one-on-one conversations that pique their curiosity and motivate them to pursue greater mathematical understanding.

But it is also important that we confer with *all* students periodically, even if they do not appear to need extra support or additional challenge. The process of "talking math" with them, face-to-face, builds our relationships with them and helps ensure that they are engaged in the learning process. We demonstrate our respect for their thinking and learning as we confer. Very importantly, during conferences, we show students that we care—about them as people and about their learning. We give them opportunities to practice their mathematical communication skills in a meaningful context. And in doing this, we may well increase their mathematical interests and further their insights into the complexities of the discipline.

What Is the Structure of a Guided Math Conference?

The conference structure below was adapted from the architecture for a writing conference described by Calkins, Hartman, and White (2005). It is as effective for math conferences as it is for writing conferences, serving to establish a *predictable* conversational pattern that not only supports growth in mathematical proficiency, but also allows conversations about students' mathematical work to flow easily because both teachers and students have a good idea about how the discussion is going to go (Anderson 2000). The sequential steps of the math conference structure are:

1. Research student understanding and skill.

2. Decide what is needed.

3. Teach to student needs.

4. Link to the future.

Research Student Understanding and Skill

"The goal of a conference is to move a student from what he or she can *almost* do independently to what he or she *can* do independently" (Sammons 2010, 213). This fine distinction is not often apparent in the written work students produce. Only when we talk with students, question them about their thinking, share our own mathematical thinking, and most importantly, listen to them, can we hope to make this distinction, and in making the distinction, determine exactly what we think they need to be able to do it independently.

During the research phase, the teacher and student each have a unique role. Figure 3.1 below highlights the key features of each role.

Figure 3.1 Research Phase of a Guided Math Conference

The Role of the Teacher	The Role of the Student (The Lead Role)
· Observes student work closely · Listens carefully to student's description of his or her work · Searches for evidence of strengths and needs · Questions to clearly understand student thinking	· Shows his or her mathematical work · Explains his or her mathematical thinking · Considers and describes possible alternative strategies · Makes mathematical connections · Shares any confusion or questions about the mathematics

The Role of the Teacher

The teacher's role during the research phase is just that: research. In order to move students ahead mathematically, we must know what it is they understand and what they can do with accuracy. Only with this information can we determine and then meet their instructional needs—whether they need remediation, additional mathematical strategies, focused practice, motivation, confidence, mathematical misconceptions corrected, or mathematical thinking challenged so they think more deeply and critically.

The most important tasks for teachers during this phase of a Guided Math conference are closely observing student work and actively listening to students as they discuss their thoughts and strategies. Although the research phase is quite brief, paradoxically, it is a time for teachers to slow down (Sammons 2010). It may be hard to do, but we learn the most when we let our students take the lead here.

Scanning students' work during the research phase is important, whether they are working with manipulatives, solving problems, playing mathematical games, representing mathematical ideas, or recording their mathematical thinking in math journals. As students share their thinking, familiarity with their work makes it possible to directly connect their mathematical reasoning with the work product itself. Asking broad general questions leads learners to describe their reasoning with greater precision and depth. Specific, probing questions help us delve into areas of understanding that may not be apparent from either observation or student explanations. Listening *deeply* to their accounts of their thinking allows us to glimpse the work through the students' own lenses (Fletcher and Portalupi 2001).

Drawing upon both observable work products and students' verbal explanations, teachers search for evidence of students' strengths and needs. During some conferences, we may find that the overall picture students provide for us is hazy and incomplete. When our students struggle to clearly describe their thinking, we need to continue to pose questions that require them to make their thinking more transparent. With these questions, we gently nudge students to say more about their thinking and their work (Anderson 2000).

Will the research findings always be absolutely on target? No, probably not. But this process of discovery gives solid evidence of proficiency and understanding, or lack thereof, which teachers can use to determine the most appropriate next steps in learning for students. According to Calkins, Hartman, and White (2005), it is crucial that teachers are aware that the research phase will always be inadequate. Since the research phase should never last for more than a third of the conference, its brevity precludes a comprehensive assessment of a student's proficiency.

With just a few minutes of observation and conversation, how can teachers be confident that their assessment of student strengths and needs is on target? It is reassuring to remember that in addition to the evidence gathered in the conference itself, teachers can reflect on what they already know about a student. When conferring, there may not be time to refer to anecdotal notes or other assessment data, but being aware of students' past mathematical achievements provides a reference point from which growth in mathematical understanding may be measured.

Teachers should also keep in mind that the teaching point presented in any conference is only one minor instructional element in each student's learning. It may have great impact because of its timeliness and alignment to the curriculum and students' needs, but if the focus of the conference fails to precisely target student needs, there are ample opportunities to revise and follow-up with additional conferences or small-group lessons that will more accurately meet those needs.

The Role of the Student

The research phase of a Guided Math conference is primarily an assessment tool. It helps teachers understand student thinking, particularly regarding the mathematics relevant to their current work. To more precisely target learning needs, teachers need to know more about the thinking behind the written work of their students. So, to shed more light on their reasoning, students assume the lead role in this stage of the conference. The focus is on the student—on both his or her work and dialogue.

Although students will be called upon to share their ideas, teachers may choose to observe a student for a moment prior to beginning a conference conversation to learn more about how a student approaches his or her mathematical work (Calkins, Hartman, and White 2005). What mathematical understanding is obvious from his or her actions? Does the student appear to have any misconceptions? What strategies does the student employ? Is she or he actively engaged and eager or frustrated and overwhelmed? Reading the student's body language and other nonverbal communication often suggests an opening for the conference conversation.

Following the brief period of observation, the teacher and student interaction increases. According to Anderson (2000, 20) when describing writing conferences, "The conference begins with students in the lead role, setting the agenda for the conversation by describing the work they are doing as writers." In the case of math conferences, of course, students describe their work as mathematicians. Only when these highly personal and unique insights are shared with us can we help students become better mathematicians—guiding their development by melding what we know about them as fledging mathematicians with what we know about the discipline of mathematics.

Because the conversational pattern is predictable, students know they will begin the conference by sharing what they are thinking and providing justification for their mathematical work. Many students come to value this collaboration with their teachers. As they work on their assigned tasks, they think of how they might describe their work, consider alternative strategies to discuss, become aware of mathematical connections to share, and assess how well they understand the math concepts with which they are working—knowing that they will be expected to talk about these during a conference. So in effect, the expectation of conferring with a teacher actually shapes the mathematical thinking of students as they work.

Getting the Most Out of the Research Phase

When conferring with students, it helps to keep in mind the goal of the research phase of a math conference: learning enough to discern what the young mathematician is trying to do, is able to do, and not quite doing (Calkins, Hartman, and White 2005). Where exactly is the student in the progression from novice to expert in regard to the mathematics at hand?

With practice, teachers become more adept at both their observations of students as they work and their use of questions as they converse with young learners.

Observing Students as They Work

Stepping back to watch and listen to students as they work gives teachers a great deal of information about students' attitudes toward math, their comprehension of mathematical concepts, and the strategies they use to increase their understanding or solve problems. This is often a very productive first step in the research phase. To get a true sense of students' capabilities, it is important to focus on an individual student and then watch him or her work uninterrupted for a few minutes. While observing, it helps to reflect on the recent mathematical performance of the student. Questions about the student's work that occur during this reflection may well be answered through observation.

At times, students will make errors as they work. Although misconceptions and mistakes must ultimately be addressed, it is wise to allow time for students to recognize and correct their own errors. Much student learning results from this self-correction process. Waiting to step in and make corrections also sends a clear message to students that they are expected to assume responsibility for monitoring the reasonableness of their work, rather than relying solely on their teachers.

Watching students as they work frequently inspires teachers to wonder about particular aspects of their students' thinking or actions. These are productive topics to explore with questions following a brief period of observation during the research phase.

Questioning to Reveal Students' Mathematical Reasoning

To a large extent, students set the agenda for this phase of a math conference because its success hinges on finding out exactly what mathematical concepts and strategies are shaping students' work. When they set the agenda, they express their mathematical ideas in a way that makes sense to them rather than in a particular order dictated by a series of questions from the teacher. It reveals much more about students' reasoning and strategic competence.

Teachers invite students to set the agenda by asking them open-ended questions that prompt them to share their mathematical thinking. The repetitive use of a predictable opening question signals students that their role in the conference has begun and that it is an important role. There is not necessarily one "best" question with which to begin, but when one works well, it is helpful to use it regularly to begin conferences. As Anderson (2000, 29) explains, "By using a predictable opening, I'm simply taking advantage of students' implicit knowledge of the nature of conversations, and that some conversations begin in a predictable way." Then, students are clearly aware that it is their responsibility to take the lead role.

Open-ended research questions also give students the encouragement many of them need to answer honestly, without fear of displeasing their teachers (Anderson 2000). These questions do, in fact, lead students, but lead them toward an understanding of the way mathematicians think about their work.

The questions below are examples of open-ended questions that teachers may use to initiate math conferences with students, but the list is not an exhaustive list. Teachers should find and then use questions that work best for them.

- *How's it going with your work today?*

- *Will you tell me what you are working on?*

- *What mathematical strategies are you using today?*

- *What are you doing today as a mathematician?*

- *How do you feel about your work today? Why?*

As students respond to the first question during a conference, their responses often call for follow-up questions. Again, the follow-up questions should be as open-ended as possible while still spurring students to more explicitly explain their thinking.

- *What is your plan for your work today? Is it working for you? Why or why not? Have you thought of any other strategies you might use?*

- *Why did you decide to...?*

- *What do you think are the most important aspects of the math you are doing?*

- *What predictions can you make based on your work?*

- *What questions could you ask that would help you to understand the math better?*

- *What do you wonder about regarding the mathematics we are studying?*

If a teacher is unsure of how to prompt students to expand on their explanations, sometimes the best strategy is to simply describe what he or she has noticed about a student's work (Calkins, Hartman, and White 2005). If a student has made little progress with his or her work and appears to be stumped about how to proceed, rather than asking what his or her plan is, the teacher may choose to say:

- *I see you have drawn quite a few different representations of the problem you are working on, but have not used any of them to find the solution to the problem. Can you tell me about your thinking on this work?*

- *You seem to be (stumped/frustrated/at a standstill). Can you tell me why?*

Especially when working with mathematics, it is tempting to ask questions with known answers during conferences. However, responses to these questions tend to inhibit conversation. The students' responses to these queries may be correct or incorrect, but they seldom lift the veil on their mathematical reasoning or strategic thinking. Instead, authentic and honest questions to which teachers really want to learn the answers invite students to more fully describe their thinking. Teachers may use these stems and questions to explore students' reasoning:

- *Why do you think…?*

- *What do think would happen if…?*

- *What mathematical connections were you making when you…?*

- *What do you think was the most effective math strategy you used to solve the problem? Why?*

- *Are there other ways of expressing/representing this?*

If teachers notice a student who is not working productively, they may decide to confer with him or her by taking one of two approaches (Calkins, Hartman, and White 2005). The first is to begin the conference as if the student were working diligently by asking a question such as, *What are you doing today as a mathematician?* With this kind of prompting, students may be motivated to begin thinking mathematically and delve into their assigned tasks when the conference concludes. On the other hand, sometimes the best approach is the more direct one. Teachers may choose to simply share their observations of the student's lack of productive work and their expectations of mathematical engagement in the future with these students and then continue to monitor their work. Either of these approaches offers students opportunities to share any confusion they may have about the assignment or other reasons why they may be stumped or frustrated by the math task.

Challenges with Conference Research

As informative as the research phase of math conferences may be, problems sometimes arise. Common challenges teachers encounter during their one-on-one math conferences include:

- **Too much of the conference time is spent on the research phase** (Calkins, Hartman, and White 2005).

  Teachers who are conferring with students for the first time may devote excessive amounts of time to research simply because there is so much that they can learn by speaking directly with them. Each conference becomes so lengthy that very few can be conducted. Teachers begin to feel that the entire process of conferring is unmanageable. As a general rule, the research phase should consume no more than one-third of the conference time. With experience, most teachers discover an effective balance of time between the phases of a math conference and are able to focus on unveiling the most pressing needs of their students in an efficient manner. Gradually, the research process becomes almost second nature for them. They allow students ample time to talk, but they actively listen and keep the discussion focused on furthering their knowledge of student understanding (Sammons 2010).

- **Teachers try to correct student misconceptions and extend their thinking with multiple teaching points during the research phase**.

  It is tempting to launch into teaching immediately upon discovering a misconception or student need. As teachers, we want to teach. It makes us uncomfortable to listen as students share a misconception in their thinking or show us work with obvious errors. Experienced teachers, however, recognize the value of gathering and weighing all evidence prior to deciding what to teach. Only by making full use of the research phase can teachers most accurately identify how well students understand the math with which they are working and establish a sense of collaboration with students. Moreover, while there may be a number of possible teaching points that eventually emerge from the research phase, math conferences are most effective when teachers first consider carefully overall student needs as evidenced during this phase of the conference and then limit their one-on-one teaching to a single, specific point. Additional student misconceptions or needs should be noted and addressed in another conference, during small-group lessons, or even during large-group instruction, if appropriate.

- **Research information is not used to determine the next steps in learning** (Calkins, Hartman, and White 2005).

  This problem is by far the most serious of the three. While the research phase may not clearly indicate which of several teaching points is most pressing, it is the strong link between the research findings and the choice of a teaching point that makes the math conference such an effective teaching tool. Its precision is contingent not only on the quality of the research, but the direct relationship between the two. As *Guided Math: A Framework for Mathematics Instruction* warns, "When the lack of connection between research and teaching point occurs on a regular basis, ... it indicates a failure to comprehend the purpose of conferences" (Sammons 2010, 216). The impact of the short, individualized lessons that are characteristic of a math conference is wholly dependent on the fidelity of the lessons to the needs that have been identified.

Decide What Is Needed

The decision phase of a conference builds upon what a teacher discovers during the research phase and frames the focus of the math conference. This pivotal point of the conference, which is largely responsible for its instructional value, occurs almost simultaneously with the research phase (Sammons 2010). Teachers assume the lead at this point, while students move into a more passive role. Weighing what they have learned about their students' strengths and needs and what they know about their mathematics curriculum, teachers decide how to proceed with the conference. The distinct roles of the teacher and the student are summarized in Figure 3.2.

Figure 3.2 Decision Phase of a Guided Math Conference

The Role of the Teacher (The Lead Role)	The Role of the Student
· Decide what the student has done well and offer an authentic compliment · Decide on a teaching point · Decide how to teach the teaching point	· Accept and reflect on the compliment of the teacher · Remain mathematically focused · Prepare to attend to the teaching point

The Role of the Teacher

Throughout the research phase, teachers are assessing progress of students toward their mathematical learning goals. Once they gain a feel for a student's degree of conceptual understanding and strategic competence, they must then rely on their professional judgment to decide how best to move the student forward from his or her present understanding and capability in order to meet or exceed learning expectations.

Deciding on an Authentic Compliment

The first decision teachers make during the decision phase flows from their observations of what a student understands and has done well. Calkins, Hartman, and White (2005, 64) advise writing teachers to consider the question, "What has this writer done well that I can compliment and therefore reinforce?" The same advice applies to math teachers.

This is not always an easy task because the compliment must be *authentic*—both in the sense that it is of value mathematically and that it is indeed something the student has done. It is important for students to know precisely what it is they have done well. When students are really struggling, teachers may have to focus on very small steps toward mathematical growth, even attending to the partially correct. Johnston (2004, 13) suggests that noticing "*first* the part of the student's work that is correct is a perceptual bias we need to extend to students." By confirming the worth of these aspects of student thinking or work, students are encouraged to repeat them.

Positive recognition also affirms the learner's competence, giving him or her the confidence to continue trying. Teachers may compliment a student on writing numbers legibly so that they can be read without errors—just like mathematicians do! Or perhaps compliment a young mathematician for getting out manipulatives as a first step in solving a problem. On the other hand, a teacher may see quite a few things that a student has done well that might lead to a compliment. Then, he or she has to decide which of several possible compliments will be most valuable in promoting the mathematical achievement of the student.

Giving students positive feedback in the form of an *authentic* compliment on their work serves two functions. First, the recognition of strengths establishes an encouraging tone and prevents the conference from seeming overly critical. Students feel supported and proud of their work when teachers notice and affirm their mathematical growth. They are more receptive to the ensuing teaching point as a consequence. Students who lack confidence in their mathematical abilities, and who feel vulnerable when teachers approach them to confer, especially value compliments. We want to challenge students with mathematical assignments that push them to go beyond their comfort levels, to try new strategies that may or may not work for them. Compliments serve as not only confirmation that they are learning, but also spotlight things that they should continue to do (Sammons 2010). Second, even when students are doing something well, they may not be aware of exactly what it is they are doing well. Once it is explicitly pointed out to them with a compliment, however, it becomes something they will try to replicate in the future. As such, the compliment may serve as an entry to the teaching point—a way of illuminating what the student has done in order to extend it.

Based on what students reveal in the research phase of a math conference, the scope of possible topics to be considered for compliments is almost unlimited. Compliments may address student attitudes—perseverance, flexibility, or collaboration with others. They might target mathematical strategies used, processes applied, or computational competency. They may highlight students' mathematical communication, either oral or written, with an emphasis on the use of mathematical vocabulary or their explanations of their mathematical reasoning. Additionally, teachers may compliment the mathematical connections students make as they work or their multiple representations of the mathematics. The focus of a compliment is not confined to the mathematics currently being studied. Instead, teachers have the flexibility to reinforce good mathematical practices that they notice and that they believe will be of most benefit to their students in the future.

It is important to remember the enormous impact our conference conversations can have on learners. In essence, when they share their mathematical work with us, they put themselves on the line—hoping to live up to both their own and our expectations. When the search for things done well is an integral part of the conference, it affects the whole tenor of the teacher–student exchange.

> When we approach conferences knowing that our immediate response after the research phase of a conference will always be to give the child a long, meaty, powerful compliment, then we read the child's work thinking, "What can I gush over?" This transforms our body language, our mind-set—and above all, our relationship with the writer. (Calkins, Hartman, and White 2005, 64)

Deciding on a Teaching Point

Once teachers decide on a compliment, the next task is to determine a teaching point. The more familiar a teacher is with the mathematics curriculum, the easier this is to do. Experienced teachers know how to break the big ideas of mathematics down into smaller teaching points—that may or may not build upon each other in a sequential fashion. For new teachers or for teachers working with a new curriculum, the process requires considerable thought (Sammons 2010). They weigh what they have learned about a student's work during research in light of the student's learning goals to plan an appropriate and concise "next step" in learning. At best, the teaching points of math conferences target strategies or increased understanding that will remain with students as tools that they can apply as they engage in mathematical thinking and problem solving—learning which students can use not only immediately, but which they can also draw upon for the rest of their lives. In effect, they learn the strategies and techniques that good mathematicians use.

It may be tempting at times to teach to the present work of students—a quick and easy fix to help students find a correct answer. There is a clear distinction, however, between helping students get a correct answer and helping students truly understand the mathematics that results in correct answers. As Calkins (2000, 228) explains in *The Art of Teaching Writing*, "our decisions must be guided by 'what might help this *writer*' rather than 'what might help this *writing*.' If this piece of writing gets better but the writer has learned nothing that will help him or her on another day on another piece, then the conference was a waste of everyone's time." In the same vein, during Guided Math conferences, the focus should be on helping the *young mathematician* rather than on helping the current piece of *mathematical work*.

Teaching points that address only memorized procedural steps without conceptual understanding handicap students in the long run. Students rarely retain the knowledge and are unable to apply it in other related mathematical situations. Because teachers are under tremendous pressure to prepare students for success on high-stakes tests, we are often desperate to find the quickest way to "teach" mathematics so our students will succeed. Unfortunately, teaching simple procedural gimmicks leaves students lacking a real understanding of the math.

One example of a procedural shortcut is teaching the butterfly method for adding fractions with unlike denominators (see Figure 3.3). Unfortunately, this procedure is sometimes taught without the mathematics behind it. The procedure is, of course, simply a shortcut for finding a common denominator. But unless students understand the value of finding a common denominator and have the skill to do it, they are baffled when they are asked to find the sum of three fractions, all of which have different denominators.

Figure 3.3 Butterfly Method

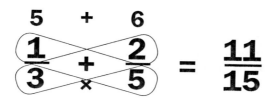

When deciding on the best teaching point during a conference, teachers should maintain an instructional focus on what is best for the long-term mathematical growth of the student, instead of aiming for lessons that provide short-term fixes.

It is reassuring to know that there is rarely, if ever, just one correct teaching point based on the research phase of a conference. Teachers have to perform a juggling act: consider the evidence of mathematical strengths and needs, reflect on the standards being taught and the learning goals of their students, take into account their learners' unique mathematical mindsets, and finally decide what teaching point can best be presented succinctly and effectively in this format—and make this decision all in a matter of minutes! It may appear at times to be a daunting task, and the choice of a teaching point may seem almost an arbitrary one. But once teachers begin conferring with their students, they discover the tremendous value of these tête-à-têtes. They realize how much students learn from teaching points that are so closely aligned with their own work and their learning needs. So they regularly confer with their students without worrying overly much about whether the teaching point they select is the one-and-only *best* one.

With so many aspects of the discipline that may become teaching points for young mathematicians, teachers have a wide range of potential teaching points from which they may choose. These suggestions offer guidance for identifying points that will be effective for promoting greater mathematical understanding and facility.

- **Teach to the young mathematician's own intentions.** If a student is trying to use a mathematical strategy but it is just not working, it is often more productive to teach him or her how to make the strategy work than to introduce and teach another strategy altogether. It is important to respect students' fledgling efforts to understand mathematics and apply their knowledge to problem solving. While their initial attempts may not always be the most efficient, they learn from their experiences, particularly when they have the chance to share their thinking with their peers. Whenever possible, teaching points should teach toward the mathematical line of thinking established by students in their work. Looking closely at student work and listening closely as students discuss their ideas will lead teachers to next steps that are aligned with students' learning progress.

- **Teach to promote mathematical independence.** Learning how to draw upon what students know and can do mathematically as they encounter new concepts and problems not only broadens their knowledge and skill but also gives them the ability and confidence to function independently in the realm of mathematics. Mathematical teaching points are most effective when they lie within what Vygotsky (1978) termed the student's zone of proximal development—something the student can *almost* do independently, and with just a little more support or practice will be able to do on his or her own. Designed to strengthen and extend the understanding and skill evidenced already in the conference, a teaching point provides the necessary scaffolding for a student to take the next step independently.

- **Teach for mathematical comprehension.** When we are mathematically proficient, we know more than just the steps for computation or the way to solve a particular problem. We have an understanding of math concepts that we can call upon when we encounter math-related tasks or situations. Our understanding is a basis for interpreting the world around us. It is what we draw upon to solve the problems we deal with daily. The better we understand mathematics, the more adept we are. Teaching points can lead students to a deeper understanding of mathematical concepts and prompt students to explore how the concepts are interrelated and are relevant in the real world. Teachers may use a math conference to review previously taught concepts, reinforce a student's understanding of concepts currently being taught, or to briefly preview a new concept.

- **Teach for strategic competence.** It is important for students to have not only a toolbox of possible mathematical strategies available to them, but also an awareness of when and how these strategies may be used effectively. Students may be reminded of strategies that have been taught and practiced earlier in the school year, or even in a previous grade. The teaching point may address how a strategy applied in one context may be used in a different context. Or, the lesson may focus on how to determine what strategies could be applied to solve a particular problem. If a student has chosen a fitting strategy to pursue, he or she may need some instruction to fine-tune the application of the strategy to the problem.

- **Teach to foster mathematical communication.** The language of mathematics is not one that most students have experienced before they come to school. Expressing mathematical ideas and sharing their own mathematical thinking is new for many of them. And so the teaching point of a math conference may focus on the use of mathematical vocabulary, what communicating mathematically really means, or on how a student can more clearly justify his or her mathematical thinking or work. Teachers can both model the proper use of mathematical terminology and prompt students to more precisely express their own thinking in a mathematical manner as a teaching point during a math conference.

- **Teach to create a community of young mathematicians.** While this is a somewhat broader and perhaps less content-specific topic for a teaching point, it is one that can have an enormous impact on the climate of a classroom. Creating a community of mathematicians leads to a classroom environment in which students feel comfortable extending their wings to try new approaches to problem solving and thinking beyond what they have been taught. This kind of community develops students who are curious and willing to make mistakes as they make conjectures, question one another's reasoning, and truly act as practicing mathematicians. The teaching points of a conference can serve to nurture a sense of community. Teachers can help students learn how to respectfully question the thinking of others, how to learn from their own mistakes and the mistakes of others, and how to consider and learn from the ideas of their peers.

Determining How to Teach the Teaching Point

After deciding what to teach during a math conference, a teacher has to determine the most effective way to present the teaching point to the young mathematician with whom he or she is conferring. Since the time available for teaching is limited, deciding on the best method of teaching is important. Simply telling a student what he or she should know does little good. Students must come away from the conference knowing and/or being able to do something that extends beyond their previous understanding or skill. During the decision phase, teachers plan how to best deliver the instruction. Making that decision may be challenging, especially for novice teachers or teachers who primarily use whole-group instruction.

Drawing on the transcriptions of hundreds of writing conferences, Calkins, Hartman, and White (2005) identified the teaching methods most frequently used by writing teachers during one-on-one conferences:

- Guided practice
- Demonstration
- Explanation of examples

Any of these methods may be used for teaching mathematics, as well. Teachers may choose to use any of these three instructional options as they plan how to share teaching points with their students. Each of these methods gives teachers a way to concisely present a brief lesson while actively involving the student. These methods will be elaborated upon in the following section on the teaching phase of Guided Math conferences.

The Role of the Student

The decision phase lasts only a few seconds (often occurring simultaneously with the research phase), so the student's role is limited yet critical to the success of the conference. As the teacher assumes the more active role, the student listens attentively and reflects. How does the compliment correspond to his or her own assessment of the work and ideas shared? This active student reflection is valuable preparation for the teaching point of the conference. When teachers first begin conferring with students, it is helpful for them to share their expectations for students. With explicit guidance, students learn to listen attentively and carefully consider feedback.

Challenges with Conference Decisions

Although the decision phase of the conference occurs within a matter of minutes at most, there are some problems that commonly occur.

- **Compliments are not authentic, are not aligned to the mathematical goals of the student, or are not offered at all.** Compliments are an integral part of math conferences. If a compliment is not authentic—it does not accurately reflect the work or ideas of the student—it is of little value. Students know what *they* have done and thought during their math work. It is readily apparent when the compliment does not address either of these. If the compliment is authentic but does not align with the mathematics learning goals of the student, its value is also diminished. Compliments that connect what the student is thinking or has done with specific goals are most effective. And finally, when the compliment is completely omitted, teachers fail to set the stage for the teaching point. An authentic compliment establishes a climate in which a student is receptive to learning how to take the next step in learning. If well-presented, it also inspires students to wonder about those next steps. Without the compliment, a student may feel the lesson the teacher shares is an indirect criticism of his or her work. As a result, he or she may be less receptive to it.

- **Teaching points are not based on the research.** One of the most common mistakes in the decision phase is choosing a teaching point based on a preconceived idea of what a student may need rather than addressing specific needs discovered during the research phase. No doubt, teachers are aware of many student needs prior to a conference. These are certainly areas that should be tackled instructionally at some time. But for teaching points to be most effective, students must see the link between the teaching and the mathematical conversation they have just had with the teacher.

- **Teachers begin instruction without first deciding how to teach the teaching point.** The teaching in a math conference is of necessity somewhat spontaneous. Within a matter of minutes, teachers have to determine what the teaching point will be and then actually teach it. In spite of this short time frame, it is still crucial that teachers decide how to teach the lesson before they dive into teaching. Although teachers inexperienced with this format may take a little more time considering how to best share the teaching point, with practice, teachers are able to fairly quickly match the next instructional steps for a student to a method of teaching that will work best.

Teach to Student Needs

The teaching phase occurs when teachers implement the decisions they have made earlier in the conference. Having a good idea of the mathematical thinking and the needs of the students with whom they are conferring, teachers carry out the teaching plan they created during the decision phase. The lessons target specific student needs with one-on-one instruction. Teachers teach a short lesson, scaffolding the learning of their students. The scaffolding afforded the student is flexibly adjusted according to the teacher's observations of the student. It is through these lessons that students' learning is extended and enhanced, building precisely on their existing knowledge and skills. The roles of the teacher and the student are outlined in Figure 3.4 below.

Figure 3.4 Teaching Phase of Guided Math Conferences

The Role of the Teacher (Shared Role)	The Role of the Student (Shared Role)
· Teach the teaching point · Monitor and assess student understanding of the teaching point · Provide scaffolding, if needed, to ensure student proficiency	· Practice what is being taught · Explain the teaching point in his or her own words · Monitor his or her understanding and share any lack of understanding with the teacher

The Role of the Teacher

Moving into this phase of the conference, the teacher has several responsibilities. He or she may give some specific feedback to students about the validity and quality of the student's mathematical work (from which the teaching point is derived), teach a brief lesson to address a student need, and then "nudge students to have-a-go" (Anderson 2000, 56). This guided practice both helps the student internalize the teaching point and provides the teacher with an idea of exactly how well the learner has understood and can apply the new knowledge or procedures.

After giving feedback on the student's work, the role of the teacher is to share a concept, strategy, procedure, or some other mathematical insight that takes the student to his or her next step in learning. The decision on what and how has already been made, and at this point, the teacher is carrying out the lesson and monitoring the understanding of the student. The instructional strategy employed may be one of the three listed in the previous section on the decision phase: guided practice, demonstration, or explanation and modeling.

Teaching through Guided Practice

Research shows that students learn best when they are actively engaged. With guided practice, the teacher provides instruction on a concept or strategy, and then serves as a coach while the student gives it a try. This ensures that the student immediately applies the new approach or strategy to the assigned mathematical task. The teacher is by the student's side to support these first tries. The guided-practice approach allows teachers to scaffold the next steps in mathematical learning for students (Sammons 2010). It is not enough, however, to just maneuver a student into taking these mathematical steps. For students to truly understand the math involved, it needs to be explicitly described and linked to the math task—first by the teacher, and then, after some practice, by the student.

Guided Practice Snapshot

Jon and his teacher were conferring about his approach to determining the number of canned goods his basketball team will be able to contribute to a food drive if each team member donates two cans. With the research phase, his teacher discovered that Jon is drawing representations to solve the problem. Since there are nine players on his team, he drew nine stick figures, each with a can in each hand. Jon's class has practiced skip counting by two, three, five, and ten, so his teacher decides to implement the guided-practice method to teach him how to use what he knows about skip counting to solve the problem.

After complimenting Jon for using the strategy of drawing representations to solve the problem, the teacher asks Jon what he would do to find out how many canned goods would be collected if each team member donates three cans. Jon replied that he could draw one more can in the hand of each figure. The teacher tells Jon that mathematicians think about what they know about math when they have to solve problems and can sometimes discover other strategies to use that might be more efficient. He reminds Jon that the class has been practicing skip counting. Jon glances back at his representations and has an "aha!" moment. He proudly counts by two up to eighteen. The teacher asks Jon to explain what he was doing, and Jon proudly says that he was using skip counting to solve the problem. He then makes a connection to the teacher's question. He confidently points to each stick drawing and counts by three to answer the question.

The teacher is pleased that Jon has made these connections, but wants to be sure that he understands that the strategy can be used without a picture:
Jon, can you find how many cans will be collected if each team member donates five cans, but without looking at your drawing?

Jon looks around the room for a minute, and then goes to the number line. Pointing at each number from one to nine, he counts by five and arrives at forty-five cans as a solution. The teacher asks him to explain how he found the solution. Jon explains that he used what he knows about skip counting:
It was like each number was five, so I counted by five.

Jon clearly made the connection between his ability to count by two, three, or five and the solution to his problem. Because the teacher was guiding the practice, Jon was prodded to extend the use of a new strategy to a variation of his original problem without referring to his drawing. The teacher makes a note to monitor Jon's work to be sure he begins to use this strategy independently.

Teaching through Demonstration

Teachers may choose to present the teaching point through demonstration of a strategy or mathematical thinking. This is an alternative approach in which teachers think aloud as they model what they want students to emulate. It takes advantage of the natural inclination of people to watch and then copy what others do. Even youngsters try their best to imitate the behavior of older siblings. Without being conscious of it, they learn by closely watching and then replicating. Their first efforts may be poor copies, but with practice, their skill grows. People tend to be hardwired to learn through demonstration.

When teaching through demonstration, it helps for teachers to break down the teaching point into achievable steps and describe the reasoning behind each step as they are modeled (Sammons 2010). The modeling and think-aloud process allows students to see and understand what the strategy or process should look like. Explicit explanation by teachers focuses student attention on the most important aspects of the demonstration. This prevents students from being distracted by incidental observations that add little to their understanding or skill. As these lessons end, teachers remind students of exactly what they should have noticed and should remember from the demonstration, emphasizing that students should always consider these points when they work as mathematicians.

Demonstration Snapshot

Consider the scenario in the Guided Practice Snapshot. What if the teacher had not thought Jon was going to be able to make the connection between his problem and his ability to skip count? The teacher may have decided to teach the same lesson through demonstration. The compliment the teacher offers would remain the same, but in this instance, the teacher would continue by demonstrating the process of skip counting:

When you told me how you solved this problem, another strategy for solving it occurred to me. I bet it is one you might be interested in using, too. I was thinking about all the work we have done to learn how to skip count, and I made a connection! It seems to me that when each team member donates two cans of food, that's just like having groups of two. I could skip count by two to find out how many cans there would be. I know there are nine team members—so I have to count nine times. Let's see: two, four, six, eight, ten, twelve, fourteen, sixteen, eighteen. We will have eighteen cans. I just thought of the number of cans every team member will bring and then skip counted by that number.

How many cans do you think you will collect if every team member donates 5 cans? Can you try the strategy I used?

With this question, the teacher assesses how well Jon understands and can actually use the new strategy. If he seems close to being able to apply this strategy independently, but is not quite there, the teacher may choose to model and think aloud again or lead the student through the process with carefully crafted questions. On the other hand, if he has really not yet grasped this approach, the teacher may decide to teach it again later, either in another math conference or in a small group.

Using the teaching-through-demonstration approach, the teacher shared his own mathematical thinking and problem-solving strategy with his student. The learner was able to glimpse and learn from the insights of a more experienced mathematician.

Teaching through Explanation of Examples

A third method of teaching that might be used during a Guided Math conference is the explanation of examples. Classrooms that have a rich environment of numeracy often have class-made anchor charts that explain concepts and processes in the words of the students. To create these charts, the teacher serves as a facilitator and scribe, recording the ideas of the class as they discuss math concepts, strategies, and procedures. These anchor charts are particularly meaningful to students because they have a role in creating them. As such, they are a valuable resource for explaining and showing examples of the teaching point of the conference. Focusing the student's attention on these charts not only teaches and reinforces the teaching point, but also leads students to consult the charts whenever they have questions or just need confirmation of their own mathematical thinking. Other sources of displayed examples may be graphic organizers for Problems of the Week or Problems of the Day that have been posted in the classroom. Furthermore, textbook examples may be sources of examples to be used for explaining the teaching point.

Another means of explaining and showing an example is sharing the work of a fellow student. This can be of value not only for the student in the conference, but also for the student whose work is being shown as an example. Without a doubt, the recognition of sound mathematical work motivates and increases the confidence of learners.

When teachers decide to use the work of another student as an example during a math conference, it is best for the teacher, not the student, to do the explaining. This ensures that the explanation is concise and accurate—focusing on what the student should be learning from the example. While having students share their thinking and their work is a good practice in general, the conference time frame is quite limited. Teachers can both give recognition to insightful mathematical thinking and maintain the brevity and focus of a conference by explaining work themselves.

Explanation of Examples Snapshot

Returning to Jon and his problem solving work, his teacher may have decided to address the teaching point by explaining and showing an example. The teaching point itself remains the same; only the method of delivery is different. The teacher gives Jon a compliment on his work, and then says:

I was talking about a problem very similar to this with Hannah. You might be interested in seeing how she solved it. Your strategy was very effective, but mathematicians find it helpful to consider lots of different strategies. Sometimes, some strategies work better than others.

The teacher and Jon walk over to where Hannah is working.

Hannah, do you mind if I share your work with Jon?

Hannah agrees—and is obviously pleased.

Hannah's sister is having a birthday party and is inviting four guests. Hannah is helping by making a goody bag for each of the guests. She wants to include three packs of stickers in each bag. So she needed to know how many packs of stickers she would need. Her problem was a lot like yours—four guests, three packs of stickers each, or four groups of three. She tried something a little different from what you did though. She made a connection to something else we have been learning—skip counting. Instead of drawing a picture to find the solution, she skip-counted by three. Three, six, nine, twelve! Just like that, she was able to solve her problem. Thanks, Hannah, for letting us look at your work!

At this point, the teacher challenges Jon to try the new strategy by finding out how many canned goods his team will collect if they each donate three cans. As Jon tries the new approach to solving his problem, the teacher provides scaffolding as needed and assesses his understanding. In the next couple of weeks, he will check to be sure that Jon is able to apply this strategy independently.

The Role of the Student

The student's role during the teaching phase is to be an active learner—to listen carefully, to reflect deeply, to monitor understanding, to ask clarifying questions, and to engage in small tasks related to the teaching point. Although the phase is definitely teacher-led, students shoulder the responsibility for learning. They should be able to restate the teaching point in their own words and engage in a task related to the teaching point. Teachers should clearly share this expectation with students when introducing Guided Math conferences at the beginning of the year.

Challenges with Conference Teaching

The teaching phase of Guided Math conferences gives students a new direction for moving forward with their learning. Because of the limited time available to share teaching points with students, common challenges with teaching in general are exacerbated in this setting. On the other hand, the intimate, conversational nature of a conference leads teachers to more quickly and adeptly focus on student needs and then quickly adjust instruction to meet them.

- **The teaching phase is too lengthy.** One of the defining characteristics of a math conference is its brevity. In keeping with each phase of the conference, the teaching phase must be concise and to the point. Students need an opportunity to give the teaching point a try. If teachers are too wordy or try to teach too much, either the conference becomes too long and unwieldy or students do not have sufficient time to engage with the newly taught ideas or strategies.

- **The teaching phase lacks focus and specificity.** The most effective teaching points are limited in nature and quite specific. It is easy to lose focus when discussing math with young learners; there is so much to share with them. Teaching is best in this phase when it has a laser-like focus on what the student should learn from the teaching point.

- **Teachers fail to revise their lesson to meet student needs during the teaching phase.** While it is advisable to maintain a clear focus, it is also important to remember that the focus is on what the student should know or be able to do, not on the form of the lesson itself. Murray (2004, 154) stresses the importance of responsive teaching during writing conferences, stating, "The responsive teacher has to learn a waiting game, calm but alert, waiting to take advantage of what the student says." In the case of math conferences, teachers also respond to the student actions during the teaching phase.

Throughout this phase, teachers should monitor student understanding and provide scaffolding as needed. There are times, too, when teachers will realize that they over-estimated the grasp of a student as they planned the teaching phase. In those cases, it is best to step back and rethink. Either adjust the teaching point or let it go for another day. Similarly, teachers may initially underestimate a student's capabilities, which demands modification of the teaching goal.

Link to the Future

As the conference concludes, it is imperative that students have a clear conception of what they have learned and an understanding that they are expected to use what they have learned in their future work as mathematicians. The roles of the teacher and the student during the Linking Phase are summarized in Figure 3.5 below.

Figure 3.5 Linking Phase of Guided Math Conferences

The Role of the Teacher (Shared Role)	The Role of the Student (Shared Role)
• Summarize the teaching point • Express expectation that in the future, the student will remember and use what he or she has learned	• Share reflection on what was learned

The Role of the Teacher

As conferences move into the final phase, the teacher summarizes what has been accomplished and calls upon the student to reflect and share his or her understanding of what was learned and thought about during the conference. These reflections give the teacher the opportunity to gauge the learning of the student and to correct any misconceptions that arise. Many times, students' final reflections will provide ideas for future small-group lessons.

With the expectation that students genuinely see themselves as mathematicians, teachers remind them that they are expected to use what they have learned in their future mathematical work. It might seem obvious that students are expected to continue to use what they have just learned. Too often, however, students fail to make the connection between what they have just learned and their future work with mathematics. By explicitly stating that expectation, the seed is planted. The seed is further nurtured as teachers frequently ask learners about the mathematical connections they are making and the strategies they are employing.

The Role of the Student

The link phase is a period of reflection for students. They listen as the teacher provides a summary of the teaching point in order to check their own understanding. Does what they think they learned align with the teacher's summary? If so, how can they describe their learning in their own terms? If not, what do they need to do to clarify the discrepancy? Perhaps they focused on a minor point that was discussed and then missed the main idea of the lesson. Or it may be that they misunderstood the lesson entirely.

The listening and reflecting students do during this phase has a valuable self-assessment component to it. Furthermore, it helps them view themselves as active learners and as practitioners of the discipline of mathematics. Students then share their individual reflections of the math conference with the teacher. Expecting students to first listen and then reflect during this phase encourages them to assume greater responsibility for their own mathematical learning.

Challenges of the Conference Link

- **The summary by the teacher is too long.** Summaries should be just that—summaries. The linking phase is not the place to reteach the lesson. If a teacher discovers student misconceptions, they may be addressed as long as the correction can be done succinctly. If extensive clarification of misconceptions is needed, it should be noted and then provided at another time, possibly with a small group of students who share a similar need.

- **Discrepancies between the teaching point and the student's reflection are not addressed.** At times, the reflection by the student may not indicate an understanding of the teaching point. The student may have missed the point of the lesson altogether or have misunderstood it. The reflection lets the teacher know what a student is thinking, and so serves as a valuable informal assessment. Math conferences are most effective when the teacher listens attentively as the student reflects, and responds—either immediately or with follow-up instruction at a later time—when students seem to be confused or mistaken in their understanding.

- **The student does not share mathematical reflections.** With the nature of a math conference, time is of the essence. But it is important to remember that the worth of a teacher's lesson is measured by the extent of learning by the student. The intimate format of a math conference affords teachers the chance to gauge the impact of their teaching when students reflect. Allowing students to shirk this task, either because of a lack of time or because they claim they have nothing to say, diminishes the value of a math conference.

Chapter Summary

Conducting Guided Math conferences in a way that has a predictable structure benefits both teachers and students. Teachers stay focused on each of the four phases ensuring that the function of each phase is carried out. The process of researching student strengths and needs, deciding on a compliment and a teaching point, teaching the logical "next step," and then linking that learning to students' future mathematical work is a practical framework for assessing, teaching, and building relationships. Students know what to expect because of the predictability of the conversational flow. Knowing what to expect is reassuring and alleviates much of the stress students may feel in sharing their mathematical thinking and describing their work. As a result, students are more confident and more receptive to the instruction they receive.

REVIEW AND REFLECT

1. In what ways can Guided Math conferences increase the effectiveness of your mathematics instruction?

2. Which of the four phases of a Guided Math conference do you think poses the greatest challenge for you as a teacher? Why?

3. Compare the roles of the teacher and the student in a Guided Math conference. How are they similar? How do they differ?

CHAPTER 4

Types of Guided Math Conferences

From the very beginning of each school year to its end, teachers of mathematics have the enormous privilege and challenge of guiding the mathematical growth of young learners. In those first days each year, students are an "unknown quantity" to a teacher; their learning potential is a mystery. Remnants of assessment data from years past offer some insight into their achievement. But a more complete knowledge of their intellectual spark and vitality is yet to come. It is the search for that knowledge that moves us as teachers to establish relationships of mutual trust and respect with students. These strong relationships allow teachers and students to freely share their love of learning, their curiosity, their thinking, and even their doubts. The blend of the seven components of the Guided Math framework fosters a classroom community in which this occurs naturally. Creating this type of risk-free environment where growth in mathematical competency is a goal shared by the class as a whole opens the door for students to excel and, in turn, provides the climate in which teachers can catch sight of that spark and vitality and briefly rejoice—before continuing to plan how best to move these learners on to even greater mathematical growth.

Undoubtedly, some would consider the sentiments in the previous paragraph to be overly rosy. Teachers find that many students struggle to master their grade level mathematics curriculum. Some learners lack the background knowledge they need to be successful. Some are apathetic. Many face daunting learning difficulties. In spite of these ongoing challenges, teachers still celebrate when they spur students to become more curious about the mathematical world around them, to grow intellectually in their understanding of the complexities of math, and to become, in general, more mathematically literate.

What does all this have to do with Guided Math conferences? The answer to this question lies in how we relate to our students when we confer about mathematics. It is only when students—particularly those who struggle—trust us enough to risk sharing their thinking that we can both lead them to see the wonder of mathematics and discover enough about their learning needs to be able to address them. The conference conversations help build the confidence of students while giving teachers the information we need to know what their next learning steps should be.

There are a number of different kinds of Guided Math conferences that teachers may choose to use with students throughout the school year. Each of them establishes the kind of learning environment just described, but each has a slightly different focus. Guided Math conferences may be compliment conferences, comprehension conferences, skill conferences, problem-solving conferences, student self-assessment and goal-setting conferences, or recheck conferences.

Compliment Conferences

Serravallo and Goldberg (2007) devote an entire chapter of their book *Conferring with Readers: Supporting Each Student's Growth and Independence* to compliment conferences. This type of conference is particularly beneficial when used at the beginning of the school year. In compliment conferences, teachers focus attention on what students are doing well to reinforce their strengths as mathematicians. It makes sense for teachers to let students know early on that their existing knowledge and skills have been noticed. As well as building relationships and increasing student confidence, recognition of background knowledge sends a message to students about how important it is that they reflect and draw upon what they already know when learning a new concept or solving problems.

But there are other occasions when teachers may want to "research with a lens of strength" when conferring (Serravallo and Goldberg 2007, 51). Teachers may hope to inspire students to continue using their newly emerging mathematical knowledge and strategies. Or if students seem unfocused, unwilling to take risks, or unmotivated, a compliment conference may buoy student spirits and increase their engagement in their mathematical work.

With the focus on student strengths, compliment conferences:

- set a positive tone so that students know their teachers recognize what they already know and can do;

- foster an environment where students feel comfortable taking risks as they work with mathematics throughout the year;

- motivate students to make mathematical connections and tap into their foundational knowledge;

- encourage students to use previously mastered strategies as well as try out those they are just learning; and

- prompt students to draw upon and apply what they already know rather than looking to the teacher for problem-solving processes.

Teachers consciously shift their mindsets when researching for compliment conferences. In most classrooms, teachers promptly notice and attend to any misbehavior by their students. When conducting research for a compliment conference, however, one's perspective shifts away from a focus on the negative. As Serravallo and Goldberg (2007, 51) describe it,

> ...when I decide to start with students' strengths, I force myself to adjust my radar, to put on new teacher lenses and begin looking for what students already are doing well.

Observing students as they work, teachers try to determine what learners know and what strategies or procedures they may be using. To supplement these observations, the teacher may ask students to describe what they are doing and/or thinking. The teacher notes evidence of strong and effective mathematical thinking, or if not yet strong, at least fledgling efforts at that kind of thinking. Selecting the one area of strength that the teacher believes will be most beneficial to reinforce, he or she then shares a compliment with the student based on that strength.

This simply stated compliment describes specifically what the teacher noticed. In addition to delivering the compliment, the teacher explains how important this knowledge or strategy is for students who are working as mathematicians. These compliments not only boost student confidence, but may also lead students to consider how their knowledge and skills can be applied in other mathematical situations.

Compliment conferences tend to be shorter than most math conferences, focusing most heavily on the research and decision phases. The teaching point may be omitted entirely, as existing and emerging strengths are highlighted. The link phase remains an essential component, however. Along with receiving praise for their mathematical strengths, students are reminded that those are exactly the kinds of things that mathematicians know and do. As such, they are expected to continue to draw upon those strengths whenever they work with mathematics.

Compliment Conference Snapshot: Grades K–2

In this snapshot, the teacher conducts a compliment conference to encourage a student to extend his use of a strategy he recently learned but is reluctant to apply in more challenging situations.

The teacher observes Jamal as his partner pulls handfuls of counters from a bag for Jamal to count. Jamal easily counts up to five counters, but when greater numbers of counters are taken from the bag, he seems reluctant to try to count them. He often puts them back into the bag uncounted, or if he does try, he constantly looks to his partner to affirm his efforts.

Teacher: *Jamal, how's it going with your math work?*

Jamal: *Okay, I guess.*

Teacher: *Can you tell me what you are doing?*

Jamal: *Just counting these things. Li pulls some out, and I count.*

Teacher: *Let's see. Will you tell me what you are doing as you work?*

Jamal: *Okay. Here are the things I am going to count. (Li, Jamal's partner, has pulled out four counters.) So let's see. One...two...three...four. There are four. (Jamal touched each counter as he counted them and moved them to the left, so the ones he had already counted were clearly separated from those he had not counted yet.)*

Teacher: *Why did you move the counters to the side when you counted them?*

Jamal: *I already did them. I haven't done the others yet. I still have to count them.*

Teacher: *Jamal, you are working just like a mathematician. You touched each counter when you counted it and moved it to the side. This lets you know easily which ones have been counted, and which have not. That is a good strategy to use when you count. Sometimes, it can be confusing when you have a lot of things to count. Which ones have you counted? Which ones still need to be counted? But when you touch them and move them to the side, you can easily tell which ones still need to be counted. That really helps when you have a lot of things to count!*

Jamal: *That's a good strategy!*

Teacher: *Jamal, keep using this strategy when you count objects—just like a mathematician. Why do you think that's important?*

Jamal: *So I won't get mixed up when I have to count. I move them and then I know which are which. Some I counted, and some I didn't yet. I won't count some of them two times or forget some.*

With this short compliment conference, the teacher recognized a strategy that Jamal had recently acquired—moving objects already counted to clearly separate those that have been counted from those that have not. Because Jamal seemed unsure about counting larger numbers of objects, the teacher hopes to reinforce the strategy he is using so that he will extend his use of it to counting larger sets of objects.

While a teaching point might have been included that would address this extension of the use of this strategy, the teacher felt that Jamal would work it out on his own as he gained confidence in his use of the strategy and so decided to conduct a compliment conference. The teacher will continue to observe Jamal to see if he begins to use what he knows for counting sets with more objects. If not, the strategy may be more explicitly taught in another conference or in a small-group lesson.

Compliment Conference Snapshot: Grades 3–5

It is early in the school year. The teacher is observing students to assess their mathematical knowledge and their use of problem-solving strategies. Students have been asked to show multiple ways to represent and solve this problem:

Task:

The average score for the Mustangs basketball team is 68 points. So far in this game, the team has scored 59 points. How many more points does the team have to score to reach its average score?

Montserrat quickly began work on the problem. Her first representation of the problem was a traditional subtraction algorithm: $68 - 59 = 9$. She then drew a representation composed of base ten blocks.

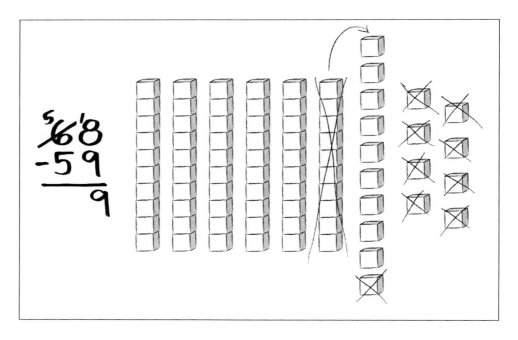

As the teacher observes, Montserrat begins to model the problem on an open number line.

Teacher: Hello, Montserrat. How's your mathematical work going?

Montserrat: Great! I know lots of ways to show this.

Teacher: Oh, will you tell me about them?

Montserrat: Okay. First, I just wrote the problem. I had to find out how many more they need to get to 68. So I made a subtraction problem. The difference was nine. So they had to score nine more points to get to the average score.

Teacher: (Pointing to the 6 that has been crossed out) What happened here?

Montserrat: I couldn't take nine from eight so I had to change a ten into ten ones. My teacher last year said sometimes we have to regroup. Right?

Teacher: Can you tell me what else you were doing?

Montserrat: Yeah. I thought about using those blocks—base-ten blocks, right? So I drew them and just showed how you can change a ten to ones. Sort of like the other problem.

Teacher: Okay. What are you working on now?

Montserrat: Well, I am going to use a number line to show how far I had to jump to get from 59 to 68.

Teacher: Montserrat, you have been thinking about lots of different strategies for solving this problem—just like mathematicians do! You knew that you were finding the difference, so you had to subtract. You wrote the subtraction problem and showed how you regrouped to find the difference. But you didn't stop there. You showed how you could draw a representation of it using base-ten blocks. Now, you are getting ready to show how you can solve it using a number line. That shows me that you know there is more than one way to solve a problem!

This year, as we work on solving problems, keep thinking of different ways to solve them. Some of those strategies will be more efficient than others, but it really helps to have several that you can use— several different ways to think about the problem.

What do you think is most important about what we have talked about today?

Montserrat: I think it is that mathematicians use lots of different ways to solve problems—just like me. It helps to think of lots of ways to solve them.

This teacher uses compliment conferences to assess student mathematical capability at the beginning of the school year and to establish positive and collaborative relationships with students. In this case, Montserrat demonstrated that she is a confident problem solver who has the ability to visualize the solution to a problem in multiple ways. The teacher reminds Montserrat that it is important to continue to apply what she knows to her future problem-solving work. Asking Montserrat to reflect on what was most important in the conference conversation leads her to verbalize what the teacher hopes she will take away from the conference.

Compliment Conference Snapshot: Grades 6–8

The students in this class will be learning about applying the order of operations when evaluating expressions involving fractions. Early informal assessments showed that Sunil and three other students were having difficulty adding fractions with unlike denominators, a prerequisite skill for being able to evaluate expressions with fractions. Their teacher worked with the four in a small group, reminding them that they must find a common denominator before they add the fractions and reteaching the process of finding a common denominator. The following day, the teacher watches Sunil practice adding fractions and decides to confer with him to be sure he understands what was taught the day before and to encourage him to continue to use what he knows about finding common denominators. Because Sunil is just mastering this skill, the teacher decides not to include a teaching point in the conference, but instead to focus on reinforcing what he is doing as he finds common denominators to solve addition problems.

Teacher: *Sunil, how is your math work going?*

Sunil: *I think it is okay. I am doing what you taught us to do.*

Teacher: *And what is that?*

Sunil: *Well, now I know that I have to make the bottom numbers the same.*

Teacher: *You are going to find a common denominator.* (The teacher reinforces the use of mathematical vocabulary by rephrasing what Sunil said.)

Sunil: *Yeah. I am finding the common denominator now.*

Teacher: *How do you do that?*

Sunil: *Well, like with this problem, $\frac{1}{3} + \frac{1}{4}$, I have to make the bottom number—oh, I mean the denominator—the same. I find a product that I can get by multiplying either three or four—the lowest one that works for both of them.*

Teacher: *The lowest common multiple?*

Sunil: *Yeah—the lowest common multiple. Four is the bigger number, so I start with it. Two times four is eight—not a multiple of three. So I try again. Three times four is twelve. It is a multiple of three. It's a multiple of both three and four, so it must be the common denominator. Okay, now it gets harder. Oh, I remember—I multiply $\frac{1}{3}$ times $\frac{4}{4}$ and $\frac{1}{4}$ by $\frac{3}{3}$, so the denominators are alike.*

Teacher: *Why can you do that? Why can you just multiply by $\frac{3}{3}$ or $\frac{4}{4}$? Won't that change the value of each addend?*

Sunil: *No, it definitely won't. I know it won't because $\frac{3}{3}$ is really one, and so is $\frac{4}{4}$. You can multiply one times any number and get the same number.*

Teacher: *You are using the identity property of multiplication to find a common denominator, just like a mathematician! You are using what you know about multiplication to find the sum of two fractions with unlike denominators. (The teacher delivers the compliment and models the use of mathematical vocabulary terms.) So, what is the sum?*

Sunil: *It's $\frac{7}{12}$!*

Teacher: *Sunil, we will be working with expressions involving fractions. Sometimes you will be adding fractions with unlike denominators. Be sure you keep using what you know about finding common denominators. What do you think was the most important mathematical idea you thought about as we just talked?*

Sunil: *I just keep remembering that I have to get the common denominator when I add those fractions, or else I won't get the right answer—I mean, sum.*

Since Sunil's ability to find a lowest common denominator when adding fractions is an emergent skill, the teacher conducts a compliment conference. This kind of math conference allows the teacher to assess Sunil's understanding of the newly acquired skill, model the use of mathematical vocabulary terms, and make sure that Sunil consistently applies what he has just learned as he works in the future.

Comprehension Conferences

Frequently during the school year, teachers confer with their students to verify learners' understanding of math concepts that are either currently being taught or have been taught previously. Comprehension of many newly introduced math concepts is contingent on the firm understanding of other related concepts students should have mastered. Because the acquisition of mathematics knowledge and skills tends to be sequential in nature, it is important that students be able to call upon what they already know to make meaning of new concepts.

Teachers can confer with students to determine the degree of their understanding of prerequisite mathematical concepts and procedures. How complete is the students' foundational understanding? Do students have knowledge of the basic prerequisite concepts that will allow them to transfer what they know to related mathematical work? How well have students retained what they were taught? Conversing with young mathematicians about their work with an eye to assessing prior knowledge reveals any lack of understanding or misconceptions.

Comprehension conferences may also be used to assess the level of understanding students have for math concepts they are presently learning. Used in this way, these conferences serve as formative assessments to shape upcoming instruction. If students show evidence of understanding as they discuss their mathematical work on assigned tasks, instruction may be needed to provide more challenge for students—leading them to a deeper, more complex understanding. If not, teachers may opt to reinforce understanding with a math conference teaching point and/or with a small-group lesson designed to strengthen conceptual comprehension.

Teachers may also use comprehension conferences to lead students to think more deeply about the mathematics with which they are working. The one-on-one nature of these conversations provides a rich setting for prompting students to think more critically. Questions such as *What would happen if...?* and *Why do you think...?* challenge students to extend their thinking and explore more complex mathematical concepts.

Comprehension Conference Snapshot: Grades K–2

Students have been learning to identify basic plane figures. The teacher observes Cara as she is sorting shapes. Using a work mat with columns for different shapes, she has put the shapes in the correct corresponding columns. In the column for triangles, the teacher notices that each triangle has been placed base-down. The teacher wonders whether Cara understands the attributes of triangles and whether she is able to identify a triangle if it is turned so that it rests on a vertex rather than a side.

Square	Circle	Triangle	Hexagon	Rectangle

Teacher: *How is your math work going today, Cara?*

Cara: *Pretty good, I think.*

Teacher: *Will you tell me what you are doing?*

Cara: *Well, I'm just sorting these shapes. It is pretty easy to do.*

Teacher: *How do you know where each shape should go on the mat?*

Cara: *I look at it and see what kind of shape it is.*

Teacher: *What do you look at to tell which shape it is?*

Cara: *I can tell by how many sides there are.*

Teacher: *Okay.* (The teacher picks up a triangle and places it on the table in front of Cara so that one of the vertices points down.) *What can you tell me about this shape?*

Cara: *Let's see—it has three sides like a triangle.*

Teacher: *Like a triangle?*

Cara: *It has three sides, but it isn't a triangle.*

Teacher: *Why not?*

Cara: *Because triangles aren't pointy on the bottom like this is.*

During the research phase of this conference, the teacher discovers that Cara can state that triangles have three sides, but cannot recognize a triangle if it does not have a side as its base. The teacher decides to help Cara focus on the important attributes of a triangle so she will be able to identify triangles no matter how they are rotated.

Teacher: *Cara, you know an important attribute of triangles! All triangles have three sides. You are thinking like a mathematician when you look at a shape and count the number of sides. But I wonder, what happens if I rotate this shape a little?* (The teacher turns the triangle so that one of its sides points downward.)

Cara: (She looks puzzled.) *Well, it has three sides, but it's not a triangle.*

Teacher: *Why not?*

Cara: *Triangles look like this.* (Cara turns the triangle so a side faces downward.) *Now it is a triangle. Triangles don't have the point at the bottom!*

Teacher: *So if you rotate a shape, it is no longer the same shape?*

Cara: *Hmm. I don't think that's right.*

Teacher: *What did you tell me you look at to decide what kind of shape each of these shapes is?*

Cara: *The number of sides it has. Oh—and they have to be straight sides, not curvy.*

Teacher: *Has the number of straight sides changed?*

Cara: *No.*

Teacher: *So why wouldn't it be a triangle anymore?*

Cara: *It just doesn't look right that way. But I guess it has to be a triangle, because it does have three straight sides.*

Teacher: *Cara, you have just discovered something about mathematics! Mathematicians have to look for the attributes of a shape—those things that make it what it is—and not get distracted by other characteristics. Would it still be a triangle if it were a different color?*

Cara: *Of course!*

Teacher: *If it were polka-dotted?*

Cara: *Yes.*

Teacher: *If it were turned?*

Cara: *I see now. I know what is important about a triangle—the number of straight sides. And the number of corners, too, right?*

Teacher: *You are right. Triangles always have three vertices, or corners. Always remember what you learned today. Whenever you are trying to identify a shape, focus on the important things that make it what it is.*

The teacher is pleased that Cara was able to express what she had learned from the teaching point of the conference. She also makes a note to be sure that students see triangles depicted in a variety of positions to help them avoid thinking that position is an essential attribute defining a plane shape.

Comprehension Conference Snapshot: Grades 3–5

The teacher observes Ruby as she compares fractions. She considers two fractions, $\frac{1}{2}$ and $\frac{1}{3}$. She writes: $\frac{1}{3} < \frac{1}{2}$. Although she properly notates the inequality, the teacher chooses to confer with her to check her understanding.

Teacher: *How is your math work going today, Ruby?*

Ruby: *Okay. I think I got the right answer.*

Teacher: *Tell me what you are doing.*

Ruby: *I was comparing these fractions.*

Teacher: *What did you find?*

Ruby: *Just like I wrote down here: one-third is less than one-half.*

So far, the teacher knows that Ruby correctly wrote and read the inequality showing the relationship between the two fractions. To be sure she really understands what she is doing, the teacher shows Ruby a number line.

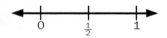

Teacher: *On this number line, will you show me where one-third is?*

Ruby places her finger between $\frac{1}{2}$ and 1.

Teacher: *Why do you think that is where one-third is?*

Ruby: *Because three is greater than two. If one-half is right here (she points to $\frac{1}{2}$), then one-third would be just a little more. So it's right about here.*

The teacher realizes that Ruby has learned a procedure for writing and reading the inequality, but does not really understand the relationship between the two fractions. Moreover, she either does not completely understand fractional parts or she is not connecting what she knows when she is asked to compare fractions. For the teaching point of this conference, the teacher decides to reinforce the meaning of fractional parts. After drawing two congruent rectangular bars— one directly above the other, the teacher continues.

Teacher: *Ruby, you wrote an inequality that was true, and you read it correctly. Those are both important skills. Now, let's think a little more about fractions. Look at these two bars. Will you please divide the top bar into halves? Tell me what you are doing as you work.*

Ruby: *I'm supposed to divide it into halves, so that means two parts.* (She draws a line correctly dividing the bar into halves.)

Teacher: *Why did you decide to draw the line there and not over here?* (The teacher points to another part of the bar.)

Ruby: *It's got to be about halfway. There has to be the same amount of space on both sides.*

Teacher: *So when a whole is divided into halves, it is divided into two equal parts?*

Ruby: *That's right, they have to be the same size—equal like that.*

Teacher: *Okay. Now, please divide the other bar into thirds. Tell me what you are doing as you work.*

Ruby: *If it is thirds, that means three parts. That's harder. They have to be equal. I'll put one line here and one here. That looks about right.* (Ruby correctly divides the bar into thirds.)

Teacher: *So now the top bar is divided into halves and the bottom bar is divided into thirds. Which is greater: one-half of the bar or one-third of the bar?*

Ruby: *The half is bigger because it is only divided into two pieces. The other bar is divided into three parts—so they are smaller.*

Teacher: *You can see that one-half is greater than one-third. Think about what you just said—that half the bar is greater because it is divided into fewer pieces—only two pieces instead of three. Knowing that, which do you think would be greater: one-third or one-fourth?*

Ruby: *Well, four is greater than three—but wait, that means the bar is divided into more pieces, so each part is smaller. I guess one-third is greater than one-fourth, right? It's sort of like dividing a candy bar. If there are only two of us, we get more than if I have to divide it between three or five people.*

Teacher: *You are thinking like a mathematician! Now, look at the number line. Show me where you think one-third would be.*

Ruby: *It's just like dividing the bar! It would be just about here, I think.* (Ruby places her finger at about one-third on the number line.)

Teacher: *How do you know?*

Ruby: *I just divided the line into three parts—equal parts—in my head.*

Teacher: *You are able to visualize thirds on the number line. You know that the denominator, the number on the bottom of a fraction, represents how many equal parts a whole is divided into. So you know that the larger the denominator, the smaller the part will be. That is something mathematicians figured out a long time ago and something that you should always remember when you are working with fractions. Will you please share something you thought about mathematically while we talked today?*

Ruby: *I just thought that it makes sense. Everybody should know it because we all share. The more people we share with, the smaller the part we get. Just like with fractions—when the bottom number—I mean, the denominator—is bigger, each part is smaller and smaller. I never knew that was math.*

With this conference, the teacher discovered that Ruby failed to connect what she knew about fractional parts to the process of comparing fractions that she had learned. She was able to express the relationship between two fractions in writing and orally, yet did not grasp how that relationship was linked to locating the fractions on a number line. By exploring what Ruby already knew, the teacher helped Ruby extend her understanding and make this connection.

Comprehension Conference Snapshot: Grades 6–8

To truly understand the structure of the number system, students must have a concept of the incredible density of real numbers. While older students regularly work with decimals and fractions, many of them fail to grasp the fact that there are an infinite number of real numbers (Van de Walle, Karp, and Bay-Williams 2010). In this conference snapshot, students were asked how many numbers they think are between five-tenths and one. While most students are busy jotting down numbers, Hakeem has written his answer: 4. The teacher decides to confer with Hakeem to help him expand his thinking.

Teacher: *Hello, Hakeem. How is your math work going today?*

Hakeem: *I'm finished already. What else do you want me to do?*

Teacher: *Will you tell me what you have done?* (The teacher chooses to take this approach rather than to tell him he needs to think about it a little more. By asking Hakeem to explain his thinking, the teacher will be able to identify a compliment and a teaching point to extend his understanding.)

Hakeem: *Okay. It's just like counting, right? So I counted six-tenths, seven-tenths, eight-tenths, nine-tenths. And next was one, so I didn't count it. What do we do next? I know that was too easy for it to be all you want us to do.*

Teacher: *Hakeem, you showed me that you understand the decimal system through the tenths. You also let me know that you understand that our work in middle school is more challenging than just counting by tenths. So now I am going to ask you to think a little more deeply about what we have been learning.*

The teacher draws a number line from 0.5 to 1 lengthwise across a sheet of paper.

Teacher: *Will you please place the numbers you just counted on this number line?*

Hakeem adds the numbers 0.6, 0.7, 0.8, and 0.9 to the number line.

Teacher: *Look at all that space between five-tenths and six-tenths. Are there any other numbers represented in that space?*

Hakeem looks puzzled.

Hakeem: *Not tenths.*

The teacher draws a number line from 0.5 to 0.6.

Teacher: *This is another way to represent that space. Is there a number you can think of that would represent this point?* (The teacher points to the line midway between 0.5 and 0.6.)

Hakeem is still confused, so the teacher points to a hundredth chart.

Teacher: *Show me where five-tenths is on the hundredth chart.*

Hakeem points to 0.50 on the chart.

Teacher: *And now, where is six-tenths?*

Hakeem points to 0.60.

Hakeem: *Ohhh! Now I get it. There's like all those hundredths between five-tenths and six-tenths—like fifty-one hundredths, fifty-two hundredths. All those numbers. There are lots of them in there between five-tenths and one.*

Teacher: *Hmm. Do you think there are any more numbers between five-tenths and fifty-one hundredths?*

Hakeem: *Of course. You gotta go to thousandths, too. But you can't stop there! How about ten-thousandths? Was this a trick question? You could just keep getting smaller and smaller parts. How could anybody figure that out?*

Teacher: *Now, Hakeem, you are thinking like a mathematician! You are wondering about the math right there in front of you. What do you think? Can anybody figure out exactly how many numbers there are between five-tenths and one?*

Hakeem: *I don't think it is possible. There are just more and more—just like infinity. It can go on and on. But doesn't it have to stop somewhere?*

Teacher: *Think about it and talk about it with your friends. After a day or two, we are going to talk about it together as a class and see what we think. This kind of thinking is what mathematicians do. Keep on wondering, Hakeem!*

With this conference, the teacher was pleased that Hakeem became so curious about the density of real numbers. Although he was focusing on decimals and had not even considered fractions, Hakeem was beginning to think mathematically. Rather than leading him to consider fractions, the teacher hopes that conversations with classmates will expose Hakeem to some additional ideas. And within a day or two, the teacher will call the class together for a Math Huddle—a focused conversation of the class mathematical community—to share ideas and possibly develop a conjecture.

Skill Conferences

Skill conferences are similar to comprehension conferences, but instead of focusing on what students know about the mathematical concepts they are learning, these conferences check on what they can do. At times, the content of comprehension and skill conferences may overlap. Students must possess conceptual understanding in order to apply it. So teachers may probe students' comprehension as they assess their skill. A conference may begin as a skill conference, but during the research phase, it may become apparent that the teaching point should address the student's conceptual understanding—which is, after all, the foundation upon which the skill is based.

Skill conferences may be used to assess prerequisite skills that are necessary for students to be successful with upcoming lessons or to monitor the acquisition of skills that are currently being taught. Much grade-level instruction assumes student competency with a wide range of previously acquired skills. If they have not mastered those skills, students struggle to learn the new content. By identifying the prerequisite skills and then assessing student capability, teachers are able to give students who lack those skills additional instruction to build their capacity. When other forms of assessment fail to clearly indicate the skill level of a student, a brief conference focused on the skill will reveal a student's competency.

As with comprehension conferences, skill conferences are a venue for valuable formative assessment as new skills are being taught. Teachers learn the individual needs of students through the conference conversations and then are able to tailor future instruction to meet those needs.

Skill Conference Snapshot: Grades K–2

Mental math strategies are important for students to perfect as they work toward computational fluency. Written assessments often show only whether students have correct answers, giving little insight into the strategies students have used to obtain their answers. Even when students are asked to explain in writing the strategies they employed, many young students are limited in their ability to effectively describe their thinking. With this conference, the teacher wants to find out which mental math strategies Layla uses when adding numbers in which the sum is between ten and twenty. Layla has successfully completed a set of addition problems, but when asked to tell how she found the sum of eight and six, she wrote, "I just know that $8 + 6 = 14$." While that may be true, the teacher hopes to find out whether she knows how to use the "make ten" strategy that will be helpful to her with future computations.

Teacher: *Layla, how is your math work going today?*

Layla: *I'm just about finished with it.*

Teacher: *I am really curious about the mental math strategies you are using as you work. Let's look at this problem: 8 + 6 = 14. Tell me what you thought about as you found the sum of those two addends.*

Layla: *Well, I just know my math facts! So it was really easy. I remember that eight plus six equals fourteen.*

Teacher: *If you didn't know that math fact, can you think of a strategy you could use to find the sum?*

Layla: *Oh, yeah! Lots of them. I could start at eight and count six more using my fingers. Like this: eight, nine, ten, eleven, twelve, thirteen, fourteen.*

Teacher: *That works. You said there are lots of strategies you could use. Will you tell me about another one?*

Layla: *I also know that six plus six equals twelve. Eight is just two more than six. So that's twelve plus two, and that's fourteen.*

Teacher: *Those are two great strategies you can use. Do you have any more in your toolbox of strategies?*

Layla: *Yep! Since eight is almost ten, I can take two from the six to make ten. Then, I just add the rest—four—to ten to get fourteen.*

The research phase has revealed that Layla has a number of mental math strategies she can use—including the "make ten" strategy. As a teaching point, the teacher decides to help Layla consider how she can apply that strategy as "make twenty" when she is adding larger numbers.

Teacher: *Layla, you can describe quite a few strategies for finding the sum of two numbers: counting on, doubles plus two, and "make ten." Mathematicians know how helpful it is to have so many strategies to call upon. Let's think, for a minute, about the "make ten" strategy to see where else we can use it. What if we are trying to find the sum of nineteen and eight? How would you find the sum?*

Layla: *I know what you are thinking! It is almost like nine plus eight, isn't it? I know that nine plus eight is seventeen. Since nineteen is ten more than nine, I can just add ten more for twenty-seven. Oh!! But I think I can also take one from the eight to make it twenty instead of nineteen. And then, seven more is twenty-seven. I see.*

Teacher: *What about eighty-seven plus 6?*

Layla: *Now, I get it. Take three from the six to make it ninety, and three more makes it ninety-three.*

Since Layla has shown such agility with mental math strategies, the teacher challenges her thinking even further.

Teacher: *Okay. Let's try one more. What is the sum of ninety-eight and five?*

Layla: *I got it! I got it! One hundred three.*

Teacher: *How do you know?*

Layla: *Easy. Two more than ninety-eight makes one hundred. Three more than that is one hundred three.*

Teacher: *Think for a minute, Layla. What do you think was the most important mathematical idea we talked about today?*

Layla: *You can use that "make ten" way to add for bigger numbers, too. I wonder if it works for adding hundreds or thousands.*

Teacher: *You can try it out and see. Let me know what you find. Maybe you can share your idea with the whole class. Remember, Layla, you have been thinking just like a mathematician. Always consider the mental math strategies you know and see how you can apply them when you are working with computation.*

By conferring with Layla, the teacher discovered that Layla easily applied the addition strategies she was using for finding sums between ten and twenty to addition problems having much greater sums. The one-on-one nature of the Guided Math conference not only allowed the teacher to accurately assess Layla's abilities, but also to challenge her to think more deeply about how they can be applied.

Skill Conference Snapshot: Grades 3–5

Marcus is trying to compute the perimeter of the rectangular class garden to find out how many feet of fencing material the class will need to enclose the area. He has measured the garden and found it has a length of twenty feet and a width of fifteen feet. The teacher observes him as he works. It appears that he has confused how to find the perimeter with how to find the area of the garden.

Teacher: *Marcus, how is your mathematics work going this morning?*

Marcus: *I'm finding out how much fence we need for the garden. I want to know how far it is all the way around it.*

Teacher: *How are you finding the perimeter of the garden?* (The teacher models using the term *perimeter* hoping that Marcus will begin using the appropriate mathematical vocabulary. If he does not begin using it after some modeling, the teacher will address its use explicitly.)

Marcus: *I measured two sides of the garden and now I am multiplying them. The perimeter is 300 feet.*

Because Marcus made it clear that he is trying to find the distance around the garden, the teacher knows that he understands what perimeter is and that finding the perimeter will let him know how much fencing is needed. The teacher's research, however, shows that Marcus has mistakenly applied the algorithm for finding area instead of the algorithm for finding perimeter. Instead of focusing on conceptual understanding, the teacher decides that the teaching point should address using the correct algorithm by referring Marcus to an anchor chart the class constructed. In addition, the teaching point will emphasize the importance of checking to be sure that answers make sense.

Teacher: *Marcus, the class has learned about finding both the perimeter and area of rectangles over the past few weeks. Your work shows me that you know what the perimeter of a rectangle is. Mathematicians know how important it is to understand mathematical terms and use them appropriately. Now, let's look at the anchor chart we made and consider how to find the perimeter of a rectangle.*

Marcus: *Uh-oh! I know what I did.*

Teacher: *What did you do?*

Marcus: *I mixed up how to find perimeter and area. Instead of multiplying, I should have doubled each side and then added them up.*

Teacher: *So what is the perimeter of our garden?*

Marcus: *Umm...Forty plus thirty, so seventy.*

Teacher: *Seventy inches?*

Marcus: No, seventy feet.

Teacher: Is there a way you could check to see whether your answer is correct?

Marcus: Yeah. I think so—if I added all the sides. Twenty plus fifteen plus twenty plus fifteen equals seventy. Oh, seventy feet.

Teacher: Marcus, you just learned something that mathematicians discovered was very important. Not only are mathematicians careful to use the correct algorithms, but also they check their work to be sure their solutions make sense. That's something I would like you to get in the habit of doing whenever you work mathematically. Marcus, what do you think is the most important thing we have talked about during this conference?

Marcus: Well, I have to remember how to find the perimeter. Duh!! It was right there on the chart.

Teacher: And how do you find the perimeter of a rectangle?

Marcus: Two times the length added to two times the width.

The teacher in this conference was able to clearly identify the cause of Marcus's error through the Guided Math conference conversation. Since Marcus clearly understood the meaning of the term *perimeter*, the teacher was able to focus the teaching point on the use of the algorithm and on the importance of determining whether an answer makes sense.

Skill Conference Snapshot: Grades 6–8

The class is working with fractions, decimals, and percent. The teacher watches Ryan work to express 7.50% as a fraction in its simplest form. Ryan writes the fraction as $\frac{750}{100}$. The teacher continues to observe to see whether Ryan catches his error. He continues to work to a final answer of $\frac{15}{2}$. To find out whether Ryan just made a careless mistake or whether he does not know how to convert the decimal to a fraction, the teacher decides to confer with him.

Teacher: *How is your math work going today, Ryan?*

Ryan: *Not too bad. I think I know how to do this pretty well.*

Teacher: *Can you share what you have done so far?*

Ryan: *Okay. I know I have to change the percent to a fraction. A percent pretty much means how many hundredths. So I just take off the percent symbol and write it over 100. Then all I have to do is simplify it. Easy!*

Teacher: *Show me exactly how you did that.*

Ryan: *Well, 7.50 percent is the same as $\frac{750}{100}$. Wait. No, it's not! I left out the decimal. I got it wrong, didn't I? Let me do it again.*

The research phase answered the teacher's question. Ryan clearly understood his error and was prepared to correct it. The teacher decides to focus the teaching point on the importance of using estimation to determine if an answer makes sense. If Ryan really understands the relationship between fractions, decimals, and percent, he should have realized that his answer was not reasonable and could have looked back to discover his error.

Teacher: *Ryan, you caught the error you made when you explained to me how you worked the problem. You know how a percent is written as a fraction. Those are both very important in mathematical work. Let's talk about something that might have helped you discover your mistake more quickly. Mathematicians learned a long time ago that we all make mistakes at times. But if we first consider what would be a reasonable answer and then compare our answer to it, it is easier to find our errors. Let's think a minute. Since we are thinking about 7.50 percent, what is a benchmark number we can use to estimate the answer?*

Ryan: *Seven and a half is greater than five, but not as big as ten. It would be easier to use ten though.*

Teacher: *So let's use ten percent and know that our answer will be less than that. If you convert ten percent to a fraction, what do you get?*

Ryan: *That's easy—one-tenth.*

Teacher: *What does that tell you about the correct answer?*

Ryan: *It has to be less than one-tenth. Oh! My answer was fifteen halves— that's a lot more than one-tenth.*

Teacher: *You are right. Mathematicians learned that it really saves time in the long run to estimate the answer first. You would have known right away that either your answer or your estimate was off. Rather than just going on to the next task, you could have corrected it. Whenever you are working on computation, remember how that can help you. Ryan, what do you think was the most important thing about math that we talked about today, and why?*

Ryan: *I guess it's that we all make mistakes—even mathematicians. But if we just think a little bit first about what answer would make sense, we can save a lot of time and trouble, too. I don't like getting math papers back with bad grades. That really gets me in trouble at home!*

In this Guided Math conference, the research phase prompted the student to discover his own error. The teacher was able to confirm Ryan's ability to convert a percent to a fraction despite the obvious initial error. The teacher could then target a broader mathematical skill with the conference teaching point—using estimation to determine whether an answer makes sense. Because Ryan had just discovered the mistake he had made, he was particularly receptive to this teaching point.

Problem-Solving Conferences

Perhaps what is most important for students of mathematics is attaining the ability to synthesize the many aspects of mathematical knowledge they have acquired to solve the kinds of real-life problems people encounter daily. Much of this ability is dependent on the toolbox of problem-solving strategies upon which a person draws as he or she taps into background knowledge and skills. In many instances, conferring with students offers a window into the thinking of students as they work to solve mathematics-related problems, making their reasoning visible. It permits teachers to understand the strategies that their students not only know, but also use to find solutions to math problems.

The Problem-Solving Process

The focus of problem-solving conferences is on both the problem-solving process and the strategies employed by young mathematicians. Polya (1957) described four steps in the process of problem solving:

1. **Understanding the problem.** A mathematician must first determine what the problem asks him or her to solve. The information available is identified and extraneous information is eliminated. During this step, it is also important to decide whether further information is needed.

2. **Devising a plan.** Drawing upon knowledge and skills, a mathematician chooses how he or she will solve the problem. In many cases, more than one approach may be used, so the mathematician decides which option is most effective and efficient.

3. **Carrying out the plan.** A mathematician then pursues the plan to solve the problem and revises it, if necessary.

4. **Looking back.** After finding a solution, a mathematician considers whether it makes sense. He or she reflects and looks back at both the problem and his or her work to assess its validity.

These steps offer students reliable guidance as they tackle any type of mathematical work. This common-sense approach clearly defines what is needed for mathematical success. Yet in their haste to arrive at an answer or solution, students tend to hurry through or skip some of these steps altogether. By conferring with students, teachers can verify that this problem-solving process is being followed and also remind students of the importance of each of these steps.

Problem-Solving Strategies

Distinct from the problem-solving process, problem-solving strategies are the methods employed by problem solvers. According to Van de Walle, Karp, and Bay-Williams (2010, 43), these strategies are "identifiable methods of approaching a task that are completely independent of the specific topic or subject matter." Students should always follow the problem-solving process, but the strategies they use will vary. Of course, the application of some strategies will be more efficient than others. Students need the opportunity to try various strategies and to share their experiences with their peers. From this kind of math talk, students learn which strategies work best, not only for different types of problems, but also for the students themselves.

Effective problem-solving strategies include, but are not limited to the following:

- **Make a guess, check it, and revise, if needed.** With this strategy, the problem solver simply makes an educated guess and then checks to see that it falls within the parameters of the problem—whether, in fact, the answer makes sense. If not, he or she gives it another go, narrowing the scope of potential solutions until the solution is found.

- **Create an organized list.** A problem solver systematically lists and considers all possible outcomes or solutions.

- **Create a table or a chart.** If there is a lot of information in a problem, it can be organized in a table or chart, making relationships and patterns more visible.

- **Draw a picture, use manipulatives, or act it out.** A problem solver may choose to use concrete objects, symbols, or even his or her own body to represent aspects of a problem. This may reveal facets of the problem that may not have been apparent at first. The use of this strategy helps the problem solver visualize and better understand the nature of the problem.

- **Look for a pattern.** This strategy may be an integral part of making an organized list, making a table, or using pictures, manipulatives, and actions. The problem solver searches for patterns in data to solve a problem. The pattern may appear in repeated items or numbers, or in series of events that occur repeatedly.

- **Simplify the problem.** By making the problem simpler—by decreasing the quantities or breaking the task into smaller parts—the problem often becomes easier to understand and manage. These insights can help a problem solver find a solution to the original problem.

- **Write an equation.** A problem solver may choose to write a number sentence as a way of organizing the information from a problem. The equation expresses the relationship between the known facts and the unknown facts of the problem, which may lead to a solution.

- **Work backward.** If a problem solver is trying to find the solution to a problem where the final outcome is known rather than the starting point, it may be helpful to begin at the end of the problem and work backward to find a solution.

Problem-Solving Conference Snapshot: Grades K–2

Li is solving the following problem:

Task:

Create a repeating color pattern with at least three colors. If you make a tower of linking cubes with this pattern, what will be the color of the 10th cube? The 20th cube? How do you know?

The teacher notices that Li has created an AABC pattern with linking cubes. To solve the problem, Li is using the cubes to build a tower with that pattern. Li seems to have the ability to create a repeating pattern and to use manipulatives to represent and solve the problem. The teacher wants to confirm his observations and possibly lead Li to represent the problem symbolically.

Teacher: *How is your math work going today, Li?*

Li: *This is fun! I am just making this pattern.*

Teacher: *Can you tell me how you are doing that?*

Li: *Yeah! I want my pattern to be red, red, blue, green. It makes it harder having the two reds first. I can put these cubes together to make the tower.*

Teacher: *Why are you doing that?*

Li: *Well, I have to find number ten. What color is it?*

Teacher: *Did you find out what color it is?*

Li: *Yep. It's red.*

Teacher: *So now, what are you planning to do next?*

Li: *I have to keep building this tower 'til I get to twenty. It will take a while—wait.*

Teacher: *I wonder if there is a way that would be a little faster?*

Li: *Nope, this works pretty good.*

The teacher's research indicates that Li knows how to create a pattern and to solve the problem by using concrete objects to represent it. As a teaching point, he decides to teach Li to use representations to extend her understanding and to help her find a more efficient method of solving the problem.

Teacher: *Li, you have worked like a mathematician and used a very effective strategy to find the answer to the first question. You used objects to represent the problem. When we were talking, though, you mentioned that it would take you a while to find what color the twentieth cube will be. What would you do if you didn't have any cubes? Can you think of a way that you could work to find the solution?*

Li: *Maybe I could use some of those tiles that we have.*

Teacher: *That is a possibility. But what if you didn't have any kind of object that you could use to make a tower? Is there any other way you could show the problem?*

Li: *Maybe we could use kids! Just line them up with different color hats or something.*

Teacher: *That would take a lot of kids, wouldn't it?*

Li: *I have an idea. What if I just used the markers and drew it instead?*

Teacher: *How would you do that?*

Li: *Well, I can draw the blocks.*

Teacher: *Can you show me what you are talking about?*

Li: *Okay. Here...* (Li draws a line of cubes with the color pattern. She frequently goes back to count how many cubes she has drawn.) *Yep! Here is number ten—it's red. I was right!*

Teacher: *How about number twenty?*

Li: (Continuing to draw) *Eighteen, nineteen, twenty! Here it is. It is green.*

Teacher: *How do you know?*

Li: *'Cause I drew the tower with the colors. I counted twenty, and it was green.*

Teacher: *So your drawing was another way of representing the problem. You didn't have to actually use the cubes, did you? Which way do you think would be best if you had to find what color the fiftieth cube will be?*

Li: *Oh, that would take too many cubes. I think I would draw that picture instead.*

Teacher: *Li, you are working just like mathematicians do! Not only can you use real objects to model the problem, you can draw a picture to solve the problem. So whenever you have to solve a problem, think about which of these strategies would work best. Will you tell me something mathematical you learned or thought about as we talked today?*

Li: *I didn't know it before, but I found out that mathematicians like to draw pictures, too. The pictures help to figure things out sometimes.*

To solve this problem, Li was clearly applying the strategy of "draw a picture, use manipulatives, or act it out." Once the teacher confirmed Li's ability to use manipulatives to solve the problem, the logical teaching point was helping her learn how to represent the problem by drawing a picture. Rather than suggest drawing, the teacher's questions led Li to discover this strategy on her own, reinforcing her confidence as an independent problem solver.

Problem-Solving Conference Snapshot: Grades 3–5

Carlos was working quickly to solve the following problem:

Task:

Michael and his father made cookies Monday afternoon. That Monday and Tuesday, his family ate 4 cookies each day. On Wednesday, there were 16 cookies left. How many cookies did he and his father bake on Monday?

The teacher watches Carlos work. Carlos writes: $16 - 4 = 12$.

Teacher: *Hello, Carlos. Looks like you are working hard. How is your math work going?*

Carlos: *Fine. I read it and found the key word.*

Teacher: *The key word?*

Carlos: *I know that when I see the word "left," it lets me know I have to subtract, right? So that's just what I did.*

Teacher: *Carlos, what does the problem ask you to find?*

Carlos: (He looks back at the problem.) *It says, "How many cookies did he and his father bake on Monday?"*

Teacher: *What does that mean?*

Carlos: *I have to find out what they had right after they cooked them—how many they cooked.*

During the research phase, the teacher discovered that Carlos is simply relying on key words rather than considering the scenario presented in the problem. In spite of that, he does know what the problem asks him to determine. She decides to help Carlos understand the entire problem, so that he can use that understanding to plan a strategy for solving it.

Teacher: *Carlos, you know exactly what you need to find to solve the problem. That is one of the most important steps in problem solving! Now, let's take a minute to consider the entire problem. Sometimes, when I have a problem to solve, I try to see it in my mind like a movie. So first, when I read "Michael and his father made cookies Monday afternoon," I can just see the two of them turning on the oven, getting out a mixing bowl, measuring out the ingredients, and mixing them up. Then, I can see the cookie sheets with the cookies on them as they go into the oven. But I just don't know how many cookies they made, do I? That's what I need to find out.*

Next, in my movie, I see his family excited about having nice warm cookies. They eat four of those cookies that day. Then, the next day, they eat another four. Finally, the problem tells us that they still have sixteen cookies.

In my movie, I see that they have a lot of cookies to start, then four are eaten, then another four are eaten. I know how many are left: sixteen cookies. Now, I just have to plan how to find out how many they had to start. I wonder if you can draw what happened.

Carlos: *Yeah, I can.*

> Carlos draws a big plate and starts to draw cookies. He suddenly stops and looks concerned.

Teacher: *What's the matter?*

Carlos: *I don't know how many to draw. I don't know how many they cooked.*

Teacher: *What do you know?*

Carlos: *I know they had sixteen at the end.*

Teacher: *Can you start there? Start with what you know.*

Carlos: *Oh, yeah. Okay, so we have sixteen cookies. (He draws sixteen cookies and then another four.)*

Teacher: *What are those?*

Carlos: *Those are the cookies they ate on Tuesday. And here are the cookies they ate Monday. (He draws another four cookies.)*

Carlos: *So they started with twenty-four cookies!*

Teacher: *You are thinking like a mathematician! What did we do that helped you find the solution to the problem?*

Carlos: *It helped when you talked about seeing it like a movie. It was like each part of the problem was like a scene from a movie. I could see it.*

Teacher: *Carlos, it is important to think about the whole problem—what's happening and what information you have. Certain words might give you some ideas about problems, but mathematicians know they have to do more than just look for key words.*

Although Carlos mistakenly relied on key words when working on this problem, the teacher was pleased to see that during the teaching point lesson he was able to work backward to find the solution. The teacher makes a note to monitor Carlos as he works on problem solving to be sure he understands the problems rather than depending on key words.

Problem-Solving Conference Snapshot: Grades 6–8

Trinity is solving the following problem:

Task:

The mean of a set of numbers is 10. What could those numbers be?

As the teacher observes her, she seems stumped and not quite sure how to proceed.

Teacher: *Hello, Trinity. How is your math work going?*

Trinity: *Not so great.*

Teacher: *What's the matter?*

Trinity: *Well, I know what "mean" means. I'm just not sure what to do to find these numbers.*

Teacher: *What does it mean?*

Trinity: *Well, it is the average. I know how to find the mean. Like—the mean of four, five, and six is five. I add them up and get fifteen, then divide by three. But I'm not really sure what to do to solve this problem.*

Teacher: *So if the mean is ten, what do you know?*

Trinity: *If you add up all the numbers and divide by the number of numbers, you get ten.*

The teacher has learned that Trinity understands what the mean of a set of numbers is. Although she understands the problem, she is not sure of a strategy she can use to solve the problem. Although Trinity could use guess and check, she may be better served by using manipulatives to come up with a series of numbers in which the mean is ten. The teacher is acutely aware of the fact that manipulatives are used less frequently in middle school. Students often come to see manipulatives as something only younger learners need. She wants to encourage her students to understand that mathematicians use concrete objects at times to clarify their understanding of mathematical concepts and to solve problems.

Teacher: *Trinity, you know what the term means. That is very important. In addition, you are taking the time to consider what it is you know and what you aren't sure about. That is something that is essential for mathematicians to do. So you are just not sure how to go about finding a set of numbers where the mean is ten. Have you thought about using some of our manipulatives to help you solve this problem? I want to show you what Samir is doing.*

The teacher shows Trinity a strategy being used by a classmate. She will concisely describe it and then encourage Trinity to try it.

Teacher: *Samir is using linking cubes to represent the problem. He has decided that there will be six numbers in his set. He made six sticks of ten cubes each—since that is the average. Then, he moved some of the cubes around. The sum of the six sticks is still the same, so the mean is the same. Now, he can list the set of numbers, and it has a mean of ten. Does that give you some ideas to work with?*

Trinity: *So I make six sticks of ten cubes?*

Teacher: *Samir made six sticks. Could you do the same thing with more or fewer sticks of cubes?*

Trinity: *Oh, sure. Let's see. I could do five sticks of ten cubes. That would work.*

Teacher: *Let's see.*

Trinity: *Okay. I have five sticks of ten. Now, I can move some of the cubes. Let's see. I have one stick with fifteen cubes, one stick with ten, one stick with five, one stick with seven, and one stick with thirteen. I wonder if the mean will still be ten?*

Teacher: *What do you think?*

Trinity: *I hope so. If I add fifteen, ten, five, seven, and thirteen, I get fifty. Then, I divide it by five. Yes! The mean is ten. But it doesn't always have to be those numbers. I could move them around in different ways, couldn't I?*

Teacher: *Give it a try. You are thinking like mathematicians have been thinking for thousands of years! What strategy did you use to solve this problem? How did it help you?*

Trinity: *I just used these cubes. I think that they help me see how to solve the problem. Do mathematicians really use things like cubes?*

Teacher: *They use all kinds of models. Remember that, Trinity. Sometimes, it helps to use objects to represent the problem you are trying to solve. That's something good mathematicians do! Now, go ahead and see if it still works when you move the cubes around in other ways.*

The teacher is pleased that Trinity demonstrated her understanding of the term *mean* and was able to use manipulatives to solve the problem. In the next week, she plans to work with her in a small group to extend her capacity so that she will be able to solve similar problems using representational models such as pictures or diagrams, and ultimately abstractly.

Student Self-Assessment and Goal-Setting Conferences

Self-regulated learning and academic achievement appear to be linked. Research suggests that "[s]tudents who set goals, make flexible plans to meet them, and monitor their progress tend to learn more and do better in school than students who do not" (Andrade 2010, 7). In contrast, those students who are less effective learners more frequently depend on others—usually their teachers or fellow classmates—for feedback on how they are progressing (Hattie and Timperley 2007).

Guided Math conferences are an excellent time for teachers to encourage students to review and clarify learning targets, assess their own progress toward those targets, and then set personal "next step" learning goals. In essence, a conference can lead students through Sadler's (1989) three conditions as rephrased by Chappuis (2009): *Where am I going? Where am I now?* and *How can I close the gap?*

Student self-assessment and goal-setting conferences address the following areas:

1. **Students review and clarify learning targets.** When students reflect on their own work, teachers can ensure that their young mathematicians have a firm grasp of exactly what the learning expectations entail. Since knowing the intended learning is essential to students' ability to both self-assess and then set goals aligned to the learning targets (Chappuis 2009; Marzano 2007), this is an important aspect of conferring. Teachers may find it necessary to restate the desired learning in language that is more specific to individual students. What makes sense to one student may well be unclear to another. Even when the learning goals are clearly understood, all students benefit from the process of reviewing them prior to assessing their own progress.

2. **Students reflect and self-assess their progress toward the learning targets.** Once the intended learning is reviewed and clarified, students are asked to reflect on their own progress toward meeting these targets. According to Marzano (2007, 38), reflection "refers to students reviewing a critical-input experience and identifying points of confusion, the level of certainty they have about content, preconceptions that were accurate and preconceptions that were inaccurate." In addition to reflecting on content, it is important for students to consider their ability to apply their mathematical conceptual understanding for computational fluency and problem solving. Research shows that asking students to monitor their learning and note areas of confusion both enhances learning and provides teachers with valuable diagnostic information (Butler and Winne 1995; Marzano 2007).

Students should not only provide their own assessment of their progress, but also include evidence to support their assessment. During the self-assessment process, students "confirm, consolidate, and integrate new knowledge" (Davies 2000, 6). Thus, its value extends beyond simple metacognition to the more challenging process of synthesis. As students reflect, they organize their thoughts, relate new learning to their existing background knowledge, use it to construct new mathematical ideas, and then share this thinking with a fellow mathematician—the teacher. When students know they are expected to engage with their teachers in this way during Guided Math conferences, it affects their overall approach to their mathematical work. The self-assessment and reflection process becomes a part of their mathematical *habits of mind*. In effect, students are encouraged to become true mathematicians rather than just learners of math content and processes.

3. **Students set specific learning goals.** After students reflect and self-assess their learning progress, they are asked what they can do to close the gap between the intended learning and their current progress. Or if students have already achieved the learning target, they are involved in the process of deciding where their learning will go from there.

The learning goals that students set with the teacher's assistance should be *hard goals*—specific and challenging goals that move students beyond their present levels of achievement (Sadler 1989). Ideally, these goals are difficult but achievable with effort. They identify the learning and include a plan of action. What will the student do? What is the time frame for accomplishment of the goals? What will be the evidence of their accomplishment? (Chappuis 2009)

Students may be asked to create a brief goal-setting plan like the one below.

Figure 4.1 Math Conference Goals Form (See Appendix B)

As students develop goals, it is important that they be able to distinguish between *learning goals* and *learning activities*. Learning goals are what the student will know or be able to do; learning activities are the actual tasks the student will do. Activities are the means by which students will accomplish the end goal (Marzano 2007). In the Guided Math Conference Goals form (Figure 4.1), the section entitled *Where I want to be* indicates the goal. The section *What I will do* specifies the learning activities students plan to complete to help them reach their goals.

Obviously, the self-assessment and goal setting completed by students will vary a great deal depending upon the age of the students. Nevertheless, even very young students should be involved in this kind of conference. They are quite capable of thinking about what they are learning, and then working with their teachers to set reasonable yet challenging learning goals. Figure 4.2 outlines the roles of students and teachers during Self-Assessment and Goal-Setting Conferences.

Figure 4.2 Self-Assessment and Goal-Setting Conferences

Conference Phase	Student Role	Teacher Role
Research	• Reflect • Describe learning target • Self-assess progress toward target • Present evidence to support self-assessment	• Introduce conference type • Listen • Clarify learning target, if needed • Guide student reflection with questions
Decision	Students do not have an active role in this phase.	Consider the validity of student's self-assessment and student's next steps in meeting learning target
Teaching	• Consider progress toward learning target to set a specific and challenging goal • Create plan for meeting goal, including activities and timetable	• Provide feedback, if needed • Facilitate student goal setting • Encourage student to set challenging, yet reasonable goal • Review plan with student (specificity, timetable, alignment with learning target)
Linking	Describe next learning steps in plan	• Emphasize importance of self-assessment and goal setting • Encourage student as mathematician to follow the plan for meeting the learning goal

Student Self-Assessment and Goal-Setting Conference Snapshot: Grades K–2

As her class began working to solve addition word problems, Karen's teacher shared the learning target with them. She let them know that they would be able to solve word problems in which they had to add three numbers using objects, drawings, and equations. She reassured the class that she did not expect them to be able to do this right away—this is their learning *target*.

The teacher now watches as Karen solves a word problem that calls for the addition of two whole numbers. With confidence, Karen reads the simple problem, writes an equation to represent it, and then solves it. This seems to be an ideal time to conduct a self-assessment and goal-setting conference with Karen.

Teacher: *Karen, I see you working hard on your math. How is your math work going?*

Karen: *Great! I read the problem and wrote this! See, I found the answer—oh, I mean, the sum.*

Teacher: *Do you remember when we began working with these addition problems? I shared a learning target with the class.*

Karen: *Yes—it's right there on the target poster. We are going to learn to add three numbers to solve a problem, right?*

Teacher: *You are right! And remember, we will be using objects, drawings, and equations. What I would like you to do now is think about what you already know and can do as you work toward that goal.*

Karen: *Well, that's easy. I can already add two numbers, like I did right here. I didn't even need to use those things (she points to some manipulatives) or draw pictures. I used to do that, but now I just write it like this. (She points to the equation she wrote.)*

Teacher: *Karen, you are thinking about what you already know and can do. When you showed me the equation you wrote, it showed me that you can do just what you said you can. So thinking about our learning target, what do you think your next goal should be? What do you think you should learn about or do next? Any ideas?*

Karen: *Hmmm...*

The teacher knows that setting incremental goals is a new idea for Karen. She supports her thinking with some questions.

Teacher: *What did you do when you were first learning to solve problems with two numbers?*

Karen: *We used little bears and things to tell stories. That was fun!*

Teacher: *That's what we did, didn't we!*

Karen: *Hey…I know what. I could tell stories and use the bears—but maybe with more of them.*

Teacher: *More of them?*

Karen: *Yeah. Instead of having some bears and then some more come, there would be some that were there, then some more could come, and then some more would come. I'd have to add three numbers then. Can I work with Amy so we can tell stories and act them out with the bears?*

Teacher: *That sounds like a good plan. What is it that you want to learn to do by working with Amy?*

Karen: *I want to learn to add three numbers for a story. Then, it would be kind of like a problem. I can add the numbers to find the answer.*

Teacher: *Karen, you are thinking like a mathematician. You know what it is that you want to learn and are planning what to do to learn it!*

Since Karen is inexperienced in setting learning goals, the teacher supported her thinking by asking her to recall activities that built her knowledge. The teacher does not ask Karen to set a timetable for meeting her goal; she was pleased that Karen easily identified her next learning step.

Student Self-Assessment and Goal-Setting Conference Snapshot: Grades 3–5

In this conference snapshot, the teacher has been reviewing his students' math journals. At the beginning of the school year, the class created an anchor chart listing expectations for journal entries. One of the learning targets for writing about mathematics was the use of appropriate mathematical vocabulary. In reading Hiro's journal, the teacher noticed that very few mathematical terms were used. He decides to meet with Hiro for a self-assessment and goal-setting conference.

Teacher: *Hello, Hiro. I would like to talk with you a bit about your math journal. Would you tell me a little about how you think about what you are going to write?*

Hiro: *Sure. Well, I think about what we worked on. I think about what we did. Then, I write it down.*

Teacher: *Okay. There is a lot to think about when writing about math, isn't there? And it is so important! I was thinking back to all the ideas the class had when we made our chart about writing in our math journals.*

Hiro: *Oh yeah, oh yeah! Sometimes, I look at that. That gives me ideas, too.*

Teacher: *That sounds like something that good mathematicians do. They consider what they are learning and then think about what they can do to learn it even better. The chart lists some of our learning targets for writing about mathematics. What do you think you do best of all the things listed on the chart?*

Hiro: *Well, when I write, I really write about math. I don't write about my friends or what I want to do after school.*

Teacher: *I noticed that in your journal. You do write about math and don't get off topic. That is a real strength. Now, look at the chart. What do you think you could work to do better?*

Hiro: *We said everyone should be able to understand what we write. But I don't know. Sometimes when we share what we wrote with a partner, they don't know what I mean. I have to tell them.*

Teacher: *That's an interesting observation, Hiro. You know, as I read your journal, I sometimes had a tough time knowing what you meant, too. But I think I know why. Look at number two on our chart.*

Hiro: *Two. Use math vocabulary whenever you can.*

Teacher: *Read your last entry. How many mathematical terms did you use?*

Hiro: *Oops.*

Teacher: *What did you notice?*

Hiro: *I didn't use any!*

Teacher: *Hiro, you are doing some really good reflection on your math writing. I think you have discovered something very important—and found a way to make your writing better. What can you do to work toward our learning targets for math journal writing?*

Hiro: *I've got to use more of those words.*

Teacher: *Which words?*

Hiro: *The words up there on the Math Wall.*

Teacher: *Can we work together to set math conference goals for you?*

Hiro: *Okay.*

Hiro, with the support of his teacher, creates the plan shown on the following page.

Teacher: *Hiro, that looks like a good plan. You are working like a mathematician—setting mathematical goals for yourself. Whenever you are writing or talking about mathematics, remember to use the terms you are learning to make your communication more precise.*

The teacher skillfully guided Hiro as he revisited his current learning targets. Hiro was able to accurately assess his progress toward his goals. Hiro was also able to identify a strength in his journal writing, and then focus on an appropriate need. Based on his self-assessment, he was able to formulate a short-term goal and learning activities to help him attain that goal. In contrast with the previous snapshot, this student sets a timetable for reaching his goal and decides how to measure his success in doing so.

```
Name        Hiro                Date        January 10

            Guided Math Conference Goals

Where I am:
I do a good job of writing about math—not other things.

Where I want to be:
I want to make my writing easier for everybody to understand by using
math words.

What I will do:
When I write, I will use the Math Word Wall to find words. I will highlight
every math word I use.

When I will complete it:
I am going to work on this for two weeks to see if everybody can
understand my journal better with these words. I will just keep doing it if it
helps.

How I will know when I am there:
I will have lots of words highlighted and people will know what I mean.
```

Student Self-Assessment and Goal-Setting Conference Snapshot: Grades 6–8

Students are working with exponents to generate equivalent numerical expressions. The teacher watches as Jeanine finds equivalent expressions for given expressions. She wants to confirm that Jeanine understands the concepts supporting the procedures she is using.

Teacher: Good morning, Jeanine. How is your math work going today?

Jeanine: It's good. My dad taught me an easy way to find these answers. I just cancel out those Xs to find the answer.

Teacher: Will you show me what you do?

Jeanine: Okay. $\frac{x^8}{x^6} = \frac{x \cdot x \cdot x \cdot x \cdot x \cdot x \cdot x \cdot x}{x \cdot x \cdot x \cdot x \cdot x \cdot x}$. Then, I just cross out six Xs on the top and six Xs on the bottom. That leaves x times x, or x^2.

Teacher: Look at your work on this problem: $\frac{x^3}{x^3}$.

Jeanine: That was so easy. You just cross them all out, so the answer is zero.

Teacher: Evaluate that expression if x equals two.

Jeanine: Okay. $2^3 = 8$. $\frac{8}{8} = 1$. Hey, something's wrong. It should be zero.

Teacher: Why do you think it should be zero?

Jeanine: Because they were all cancelled out. Nothing was left.

Teacher: Jeanine, your solution was correct when you found a value of one. You were using what you know about exponents and the value of a fraction when the numerator and the denominator are the same. But when you used the shortcut your dad taught you, your solution was different. It is very important that mathematicians understand the concepts they work with and if they use a shortcut, they have to understand how and why it works. Have you thought about the shortcut your dad taught you—why it works?

Jeanine: Not really. He just told me to do it that way.

Teacher: In our class, we know how important it is to think about what we understand and what we don't and to set learning goals. This seems like a perfect time for you to do this.

The teacher supports Jeanine as she creates the Guided Math Conference Goals form shown on the following page. The teacher will check back to monitor Jeanine's progress toward her goal.

Name _____Jeanine_____   Date _____March 15_____

Guided Math Conference Goals

Where I am:
I understand exponents.

Where I want to be:
I want to figure out how the way my dad taught me works most of the time when I have to divide with exponents.

What I will do:
I am going to try to work lots of problems like that without using the shortcut and then with the shortcut to see how they are the same or different. I might talk with my friends or teacher about it. I might talk to my dad.

When I will complete it:
I think I can figure it out in 1 week. That's my goal.

How I will know when I am there:
I will be able to figure out why I got two different answers. I will be able to tell about why the shortcut works most of the time.

Recheck Conferences

As mentioned in some of the conference snapshots, following up on the mathematical progress of students with whom teachers have conferred is often needed. While the follow-up may be through a variety of methods, including observations and written assessments, subsequent conferences offer an effective and more comprehensive way of verifying students' mathematical progress.

Recheck Conference Snapshot: Grades K–2

The teacher has conferred with Allie previously and discussed using the strategy of "doubles plus one" when adding. Allie appeared to understand the strategy during that conference, but the teacher hopes to confirm that Allie can apply the strategy with facility. Instead of simply providing the addition problems to be solved, the teacher plans to ask Allie to provide examples of problems for which the strategy may be used. Being able to do this is evidence of a deeper understanding of the strategy than merely being able to apply the strategy to given problems.

Teacher: *Hello, Allie. Last time we talked, we discussed a strategy for adding numbers—"doubles plus one." What can you tell me about it?*

Allie: *I remember. I know the doubles. They are pretty easy. So if I add numbers that are almost doubles, I add the doubles then add another one.*

Teacher: *Oh, I see. That is a good description of the strategy. Can you give me an example of when that would work?*

Allie: *Well, let's see—five plus five is ten. So five plus six is one more. It must be eleven—one more than ten.*

Teacher: *That explains your mathematical thinking very well. Do you think you can find the sum of thirty and thirty-one?* (The teacher decides to see whether Allie is able to extend her ability to use the strategy to solve a more challenging problem.)

Allie: (She writes the problem in vertical form, then finds the sum without connecting the problem to the "doubles plus one" strategy.) *Here you go. The sum is sixty-one.*

As a recheck conference, this conversation confirms Allie's ability to use the strategy for one-digit addition problems, but reveals that she did not connect it to the addition of double-digit numbers. The teacher has the option of either shifting into a skill conference to lead Allie to make the connection or to complete the conference by recognizing Allie's understanding of doubles plus one, but allowing more time and experiences with the strategy to help her develop a more extensive ability to apply the strategy. Since use of this strategy is still relatively new for Allie, the teacher opts to end the conference without providing further challenge. Allie is given more time to internalize her understanding while working with less challenging problems.

Teacher: *Think for a minute about our conversation. Please share with me something mathematical that you learned or thought about while we talked today.*

Allie: *Okay. I know how to use "doubles plus one." That can help me find the answer when I add things that are almost doubles.*

Teacher: *Thank you, Allie. You are thinking like a mathematician—thinking about strategies that make computation easier. Whenever you face addition problems, remember that "doubles plus one" is a strategy that might help.*

The teacher plans to continue to monitor Allie's progress and confer with her at a later date to determine whether she is ready to extend her use of this strategy to double-digit numbers.

Recheck Conference Snapshot: Grades 3–5

Liam is working to solve the following problem:

Task:

Mia's class is earning money to buy books for their first-grade buddy class. Their goal is to purchase 100 books for the class. The cost of 5 books is $10. How much money will they have to earn to meet their goal?

In an earlier conference, the teacher taught Liam to make a table to solve problems. Prior to that conference, Liam was using repeated addition to solve similar problems. With this recheck conference, the teacher wants to see whether Liam will apply the new strategy to solve the problem. If not, he plans to use the conference to reinforce the teaching point of the previous conference.

Teacher: *Hi, Liam! How is your math work going today?*

Liam: *I am trying to decide what to do. What do you think I should do?*

Teacher: *What are you considering?*

Liam: *Sometimes, I just keep adding. So five books plus five books is ten books. Ten dollars plus ten dollars equals twenty dollars. I know that ten books cost twenty dollars. I could just keep on adding five books at a time and find out how much we need to earn.*

Teacher: *Do you think that will work?*

Liam: *Yep. It will work, but I think there might be a better way—that might not take so long.*

Teacher: *Oh, yeah? What other strategy are you thinking about using?*

Liam: *I might be able to make one of those tables like we did before.*

Teacher: *Can you tell me a little more about what you are thinking?*

Liam: *Up at the top, I could write on one side "Number of Books" and on the other side "Number of Dollars" like this:*

Number of Books	Number of Dollars

Then, I could put "5" first and then in the next place "$10." (Liam adds those values to the columns.)

As he explains his thinking, Liam continues to use repeated addition to fill in the table he created rather than trying to identify a pattern and rule that will allow him to determine the dollar amounts for any given number of books. He fills in the table through twenty books.

Number of Books	Number of Dollars
5	$10
10	$20
15	$30
20	$40

Liam has decided to make a table to find the solution, but does not seem to know how to identify a rule for the table. For this recheck conference, the teacher decides to extend Liam's understanding of the strategy by teaching him to find the rule that will allow him to solve the problem more easily.

Teacher: *Liam, you are using a great problem-solving strategy by making a table. Let's think about how the strategy can be even more effective. Look at the table you have created. What do you notice?*

Liam: *Well... On this side, the numbers go up by five each time. On the other side, they go up by ten.*

Teacher: *So if you look at the columns, you find a pattern of increase in each row. What do you notice if you look across the rows?*

Liam: *Five plus five equals ten, and ten plus ten equals twenty. Look, that keeps going in every row!*

Teacher: *So the number of books doubled is equal to the cost of the books?*

Liam: *That's what it looks like. Maybe I could just multiply the number of books by two.*

Teacher: *Do you think that will work?*

Liam: *Let's see: two times five equals ten, two times ten equals twenty, two times fifteen equals thirty, two times twenty equals forty. It works!*

Teacher: *You discovered a rule for the pattern! Can you think of how to use the rule to find out how much it will cost to buy one hundred books?*

Liam: *Oh! If I multiply one hundred times two, it is two hundred. One hundred books must cost two hundred dollars.*

Teacher: *Liam, tell me which strategy you used to solve the problem.*

Liam: *First, I thought about it and made the table. I started out just adding them up to find the answer—how much they cost—but then I figured out a pattern. It worked every time! All I had to do was "times two" the number of books.*

Teacher: *Yes, the cost of the books was two times the number of books, wasn't it? Mathematicians always search for patterns that will help make their work more efficient. Whenever you create an input/output table, try to determine the pattern.*

Liam is just beginning to understand the use of function tables. The teacher plans to provide many more opportunities for Liam to practice. He will continue to monitor Liam's progress during small-group lessons and through additional conferences.

Recheck Conference Snapshot: Grades 6–8

Leanna struggles with understanding the division of fractions. Her common-sense approach to mathematics serves her well at times, but in other instances leads her to misconceptions. In an earlier conference, the teacher used fraction bars to help her visualize a whole number divided by a fraction. Today, the teacher observes Leanna as she divides two by one-half. Initially, Leanna records the quotient as one. The teacher decides to confer with Leanna as a way of prompting her to reflect on what she learned from the previous conference and apply it to this division problem.

Teacher: *Leanna, how is your math work going today?*

Leanna: *Okay, I guess.*

Teacher: *Can you share your thinking with me as you divide two by one-half?*

Leanna: *Well, one-half of two is one. So I guess that is the quotient I am trying to find. But I keep thinking about what we did the other day.*

Teacher: *What are you thinking?*

Leanna: *I remember that the answer wasn't what I thought it would be. It wasn't what seemed to make sense.*

Teacher: *Leanna, you are thinking like a mathematician. You are reflecting back on what you have learned and really thinking hard about what you understand and what you are still maybe a little confused about. What did we talk about during our last conference?*

Leanna: *We used the fraction bars to find the answer.*

Teacher: *Would it help you to use them now?*

Leanna: *Yeah, I think it would. Can we use them?*

Teacher: *Sure. Use the bars to show me the problem.*

Leanna: *Okay.* (Leanna sets out two bars with a value of one each, side-by-side. Then, she selects a number of bars with a value of one-half. She pauses as she considers the manipulatives. Finally, she lines up four bars with a value of one-half each below the two bars she initially set out.)

Teacher: *Will you tell me what you are doing?*

Leanna: *I started out with two whole bars. That's two. You know, I was thinking that one-half of two was one, but when we divided before, I remember that we had to find out how many of the little fractions were in the number. So I got out the halves and put them under the ones. I could put four of them under the two whole bars. So I suppose the answer is four. Is that right?*

Teacher: *What do you think?*

Leanna: *I think it is. When I divide the whole numbers, I find out how many of those there are in the number. So like there are six twos in twelve. With this problem, there are four halves in two. That must be the answer.*

Teacher: *Leanna, I like the way you thought back to what we did in our earlier conference! You described the division problem very well. You were trying to find out how many halves there are in two. The fraction bars helped you represent it. The more you work with these problems, the easier it will be for you to visualize it without using the fraction bars. Reflect for a minute on the math you were thinking about as we talked. Share what you learned with me.*

Leanna: *I know I can use the fraction bars to show how many of a fraction there are in another number. When a number is divided by a fraction, that's what I need to find.*

Teacher: *Whenever you divide a number by a fraction, remember what you did today. No matter whether you use fraction bars, try to see the problem in your mind.*

The teacher is pleased that Leanna, recognizing that her first answer was incorrect, thinks back to her earlier math conference with the teacher. Reflecting on that conversation, Leanna is able to represent the division problem using manipulatives and then determine the correct answer. She is also able to describe her thinking—both her earlier erroneous thinking and her subsequent correction.

Chapter Summary

These descriptions of the varying types of Guided Math conferences provide a source of guidance for teachers as they confer with young mathematicians. While each kind of conference has a distinct focus, their content often overlaps. Compliment conferences may highlight the comprehension or skill of students. Because problem-solving ability is dependent upon student understanding and skill, problem-solving conferences necessarily touch on those areas. The same is true for both self-assessment and goal-setting conferences, and recheck conferences.

Teachers should flexibly address student strengths and provide teaching points targeted to individual students' needs. Guided Math conferences are most valuable when teachers rely upon their existing knowledge of their students, the curriculum they are teaching, and the conversation with students during the conferences themselves. Although a teacher may begin a conference with a particular aim in mind, it is important to be willing to revise that aim based on the evidence gathered from both the student's work and dialogue during the research phase.

REVIEW AND REFLECT

1. Which kind of Guided Math conference do you think will be most challenging for you to conduct? Why? Which kind do you think will be easiest for you to conduct? Why?

2. Which kind of Guided Math conference do you think is most valuable for students? For teachers? Why?

3. What is the value of having a particular kind of conference in mind as you begin a Guided Math conference with a student?

CHAPTER 5

Implementing Conferences in the Guided Math Classroom

The rationales for one-on-one Guided Math conferences are many, yet why are conferences so rarely a part of mathematics instruction? Although there may be many reasons why teachers are hesitant to confer with their students regularly, one oft cited concern is what the other students are doing while a teacher is conferring with just one student. To converse with an individual student uninterrupted, the other students must be capable of working independently for sustained periods of time. According to Anderson (2000, 171), "fostering this kind of independence is a difficult—and critically important—classroom management challenge." In fact, successfully conferring with students one-on-one is completely dependent on the creation of an effectively managed classroom community in which students know how to remain engaged in productive mathematical tasks without interrupting the teacher.

Another reason teachers frequently give for not conferring with students about their math work is lack of time. With the teaching day so packed, when can they find time to fit in anything else? This is a real and valid concern—one that teachers face daily. Although no one can give teachers more hours in a given school day, it does help to consider the value of the many instructional components and set priorities. Which teaching strategies are most beneficial for student learning? Considering the benefits of math conferences, how can these become a regular part of instruction?

This chapter examines these two concerns and offers suggestions for making math conferences a valuable teaching tool.

Finding Time to Confer with Students

Many teachers recognize the potential benefits of one-on-one math conferences, but facing an already overwhelming instructional schedule, struggle to find time for them. It is not impossible, however, to include math conferences as a regular, and important, part of mathematics instruction. Each conference takes very little time and can be conducted during already planned student work times. Once students learn and can carry out the routines and procedures for independent work, teachers are able to meet briefly with individual students to discuss their mathematical understanding. And if teachers are flexible, these student-teacher conversations can even be fitted in during other times throughout the day.

As Students Enter the Classroom

Although teachers have many responsibilities as students enter their classrooms each day, experienced teachers organize routines to make it go as smoothly as possible. Students learn where to put their belongings, where to place notes to the teacher, where to put homework if it is to be turned in, and what work they should begin immediately. Learning these procedures not only makes it easier for students to get started on their daily work, but also frees up teachers to engage more directly with their students at this important time of the day. Greeting students as they come into the classroom sets a positive tone for the day. Well-established morning routines also make this an opportune time to conduct a Guided Math conference if a teacher has questions about mathematics work a student turned in previously.

A student may be invited to confer upon entering the classroom, either before or after stowing his or her belongings. The teacher greets the student and begins the conference with something like the following:

> *Hello, Omar! Let's talk for a minute. I am really interested in hearing about your thinking as you completed this problem. Will you tell me a little about it?*

It is important that students do not regard a conference as an indication that their work was incorrect. Care should be taken to conduct conferences with students for many reasons, one of which may be incorrect solutions. Other reasons to confer may include students not showing their work, students who use a unique or creative strategy for solving a problem, students who need encouragement, or students who are reluctant to speak in small-group or large-group settings.

When Students Are Working Independently

One of the most common times for teachers to conduct math conferences is when their students are working independently—during Math Workshop, during practice following a whole-class lesson, or during any kind of assignment which students should be able to complete successfully without teacher assistance.

In the Guided Math classroom, independent work during Math Workshop is a regular part of math instruction. Students in these classes are already well aware of the routines and procedures that support independent work in math workstations while the teacher is involved with either small-group lessons or with math conferences. Teachers in Guided Math classes may choose to confer with only a few students interspersed between small-group lessons or may opt to spend an entire class period conferring with individuals as the other students are engaged in math workstations.

Teachers who are not using Math Workshop in their classrooms can teach their students routines and procedures for independent work, which will be addressed later in this chapter. After a whole-class lesson, teachers frequently assign students practice with the newly acquired skills. While these students may not spend as much time working independently as students in Math Workshop, teachers can still plan ample opportunities for conferences as their students review and maintain skills previously mastered or work toward computational fluency.

Sometimes, teachers assign a class project, a game, or cooperative group work on which students work independently. Although this kind of assignment typically needs greater teacher involvement, teachers often find they can confer with a few students during this time if they closely monitor the needs of the rest of the class.

During Times of Transition

In most classrooms, throughout the class or day, there are times of transition that allow teachers to meet individually with students to discuss mathematical work. Teachers who think creatively make the most of these opportunities. For example, as students record their homework assignments, pack up their backpacks at the end of the day or class, select books for silent reading, or take restroom breaks, teachers can conduct a math conference with a student. If during one of these conferences a teacher discovers more time is needed, the conference may simply continue at another time. Although using times of transition is not an optimal approach, it is better than foregoing conferences altogether. The more flexible teachers are, the more opportunities for math conferences they will find.

Balancing Math Conferences with Small-Group Instruction

Math conferences—one of the seven components of the Guided Math framework—play an integral part in the assessment process on which teachers rely for identifying instructional needs. In Guided Math classrooms, students are flexibly grouped according to these needs, and appropriate lessons are provided based on data from a variety of assessment sources, formal and informal. While assessment data from formal sources such as chapter tests or benchmark tests may indicate a student's success in choosing correct answers, it typically provides little information to pinpoint the cause of errors. Finding out why students obtained the wrong, and sometimes even correct answers, is essential in meeting students' instructional needs. What gaps in learning or skills do students have? What misconceptions in their mathematical thinking exist? Which students are ready for additional challenges? When teachers confer with students, student thinking becomes more visible—to teachers and to the students themselves. Because of this, these student–teacher mathematical conversations serve as a valuable source of information revealing both students' strengths and needs. The insights gained from these brief discussions are crucial to accurate grouping of learners for small-group lessons.

If teachers are conducting small-group lessons, when do they find time to confer with students? Many Guided Math teachers confer with students during the times suggested earlier in this chapter, but there are also opportunities to confer in conjunction with small-group instruction. In these situations, teachers may confer with students:

- at the beginning of Math Workshop as students begin their math workstations, but before the first small-group lesson;

- at the conclusion of a small-group lesson by asking a student to remain for a conference; and

- by omitting one of the small-group lessons usually taught and instead using the time to confer with several students.

As professional demands on teachers increase, teachers are becoming more and more creative and flexible in order to accomplish everything that is expected of them. It is this creativity and flexibility (at which teachers excel) that allows them to find time to confer with students. After all, what is highly valued always seems to get done, somehow or other.

How Often to Confer with Students

When teachers first consider conducting math conferences with their classes, the task may seem daunting. The prospect of finding time for individual conversations with each student in the class may be overwhelming. But when this occurs, we need to reflect on the critical difference between equality and equity, and the role of each in education. Wormeli (2006, 134) writes of this important difference in his description of a fictitious classroom scenario:

> *Two students are seated at the back of the classroom. One of them is nearsighted and cannot see anything clearly that is more than a few feet away. He wears thick glasses to see long distances. The teacher asks both of them to read, record, and learn the information written in small print on the front board, on the opposite side of the room. In order to be equal, however, the teacher removes the nearsighted child's glasses and asks both students to get started. The child needing glasses squints but can't read anything on the board.*

In this case, equality clearly took precedence over equity. Whereas fairness or equity would demand that the student be permitted to use his glasses, equality calls for their removal—either that or putting the same glasses on the other student. As this scenario illustrates, what is fair or equitable is certainly not always what is equal.

With conferring, teachers should aim for equity over equality. Student interests are rarely served when teachers confer with every student equally. Instead, when teachers aim for equity, they try to promote equal outcomes for all students by responding sensitively to individual differences. Students are treated fairly when teachers consider complex arrays of information about each of their students and then teach accordingly (Van de Walle, Karp, and Bay-Williams 2010).

Even when teachers do not meet with all students at a given time "conferring with even one student sends a message to the whole class: I see you as individual learners, and I care about your thinking" (Hoffer 2012, 149). Hoffer adds, "although it is important to connect with everyone in the class, the reality is that some students need us more than others." Teachers should bear this in mind and confer with the students who need it the most. These students may not always be the struggling students. Math conferences with students who are excelling may provide learners with additional challenges. Furthermore, there are times when teachers need to know more about particular students' thinking—no matter where they happen to fall on the learning continuum.

Nevertheless, all students deserve opportunities to talk about math face-to-face with their teachers at some time or another. Keeping accurate records of student conferences ensures that every student participates in these conversations at least several times during the school year.

Once teachers realize that they do not have to confer with each of their students on a rotating basis, the question arises: what should prompt a teacher to confer with students? The following list suggests some reasons for conferences:

- To understand a student's mathematical thinking, no matter whether the student is struggling

- To give a student needed practice expressing mathematical reasoning using appropriate math terminology

- To correct a student's obvious misconception

- To give a student encouragement

- To provide a student with extra challenge

- To prompt a student to think more deeply and critically about his or her mathematical work

- To enhance the teacher-student relationship

- To assess the knowledge and/or skills of a student

- To identify and teach appropriate next steps in mathematical learning for a student

This is by no means an all-inclusive list. When teachers make conferring a regular part of their mathematics instruction, they will discover many other situations for which math conferences are beneficial.

Routines and Procedures for Independent Work

In writing about establishing reading workshop in classrooms, Serravallo and Goldberg (2007, 16) advise that "to create this kind of community—where there is a commitment to independence, an expectation of academic rigor, and a high level of productivity—there are predictable routines and structures that can be taught." Their advice also applies to creating a classroom environment conducive to conducting math conferences. Students only realize the importance of working independently without interrupting conferences when teachers explain the goals of conferring in language they can understand—letting them know exactly what they are doing and why (Hoffer 2012, NCTM 1991). Before beginning to confer with their students, teachers should explain what math conferences are and how they lead to greater learning. Students must be aware of how much interruptions disrupt and impact math conferences. Hoffer (2012, 148) exhorts teachers, "[I]f you are serious about making time for conferring, you need to set the expectations high: Do what you are supposed to do. Don't interrupt. Ever."

Of course, that is based on the assumption that students know what they are supposed to do. Making routines and procedures explicit is essential. Slaughter (2009, 45) describes how routines in life make "the difficult less difficult and more manageable." She believes that the same is true for teaching. "If we establish routines, a way things go, we not only make our job manageable but also will very likely accomplish it successfully." Successfully establishing and maintaining "a way things go" in the classroom is the result of:

- carefully planning routines and procedures that will allow the students to function independently;

- explicitly teaching the routines and procedures so that students thoroughly understand them; and

- consistently upholding the expectations for student behaviors during independent work.

Planning Routines and Procedures

The first step in creating routines and procedures for independent student work is envisioning precisely what students should be doing. Experienced teachers recognize how all aspects of the classroom environment impact independent work.

> *Environments push us into specific psychological sets; they tell us the range and kinds of behavior appropriate for where we are. The classroom environment can, in a way, tell students to care or not care about their behavior, or even suggest disruptive behavior. By the "classroom environment" we mean here more than just the physical environment, the room itself. We also mean: the seating arrangement of the class; the procedures of that room; your procedures; and the equipment used by you and the class.* (Seeman 1994, 123)

Teachers must consider what kind of work students will be doing; how the classroom will be arranged to encourage independence yet minimize disruptions; how transitions or movement around the room will occur; and plan for ways for students to handle both predictable and unpredictable potential problems. Discussing routines and procedures with other teachers provides additional perspectives and ideas of what has worked well for them. Additionally, it is important for teachers to take into account their own classroom preferences. How much student movement can they tolerate as they confer? What noise level is acceptable? Individual preferences vary and so will the routines and procedures teachers develop.

The kinds of tasks assigned to students have a significant effect on their ability to work independently, too. Assigned work should be worthwhile, engaging, and such that students can complete it on their own with a large degree of accuracy. Practice and review of previously mastered concepts or skills is not only the kind of work that students should be able to complete independently, but also helps students retain and deepen their understanding. Independent and ongoing practice often eliminates or minimizes the need for teacher-led review. Students might also be asked to engage in activities that promote computational fluency. The assigned tasks for either practice and review or for computational fluency may involve either written work, previously learned games, or work on electronic devices. Work dependent upon math skills currently being taught increases the probability that students will either need assistance or complete it with many errors, so it is best avoided if students are expected to work with no assistance from the teacher (Sammons 2010).

Just as the work students are expected to do impacts the classroom tone when students are working on their own, so does its physical environment. Are students seated at desks or at tables? Is seating assigned or do students have a choice of where to work? Will students work individually, in pairs, or in small groups? Where are materials that students may need housed? Can students access needed materials without disturbing the work of their classmates? Where will the teacher be conferring? Some teachers also take into account sensory aspects of the classroom environment—the kind of lighting (lighting from individual lamps often has a calming effect) or whether to have soft music play as students work. While there is no right or wrong way to create an environment that encourages students to work independently, it does not happen without intention. Thoughtful planning of the physical aspects of the classroom leads to an environment that encourages students to work responsibly when teachers are conferring and can greatly affect the ability of teachers to conduct math conferences.

Good routines and procedures take into account how students transition from one task to another or move around the classroom during math conferences. It is during these times that students are most easily distracted and most apt to require attention from the teacher. It helps to have clear guidelines for students about when it is permissible to move around, so they do not have to interrupt conferences to ask for permission. When students are in the process of moving, routines and procedures should make clear the behavioral expectations of the teacher. Which routes through the classroom should they take? Can they stop to speak with others as they move? Are they expected to clean up their workspaces before transitions? When planning, teachers should give careful thought to these expectations.

Planning for routines and procedures does not end with these considerations. It also involves brainstorming both predictable and unpredictable problems that might arise and thinking about how students should respond to them. Predictable problems include:

- Students are unclear about directions

- Students have trouble getting started

- Task is too difficult

- Students finish quickly

- Interpersonal difficulties

- Technology failures

- Materials/parts missing

Routines and procedures should give students guidance on what they should do if any of these problems occur, so they do not interrupt the teacher during a conference. Because these problems are predictable, teachers can plan on how to handle them before they occur, or at least before they recur frequently (Sammons 2010).

Of course, unpredictable problems may also arise. Students should know that there are times when a teacher must be interrupted (e.g., injuries, illness, dangerous situations). Discussions about what constitutes an emergency help students understand when interruption is necessary. Other unpredictable but nonemergency problems can occur that may require teachers to step back, rethink, and search for underlying causes. Sometimes teachers find it helpful to just observe as students work independently to discover how to address a problem. In searching for patterns in the ecosystem of their classrooms, teachers are often able to identify and stem these problems (Calkins, Hartman, and White 2005). At other times, teachers may choose to raise the issue of persistent problems with the class to get input in solving the problem (Anderson 2000).

Teaching Routines and Procedures

The best routines and procedures are of little use unless students completely understand them and follow them. Too often, attempts by teachers to have their students work independently fail because of a lack of adequate teaching and practicing of practical routines and procedures. Simply explaining the expectations and providing a practice period are not enough to instill the deep understanding that students need to be successful.

Research suggests that the brain receives input via three external memory systems: visual, auditory, and kinesthetic. Once the information has been received from one of these systems, the brain stores it in the same system. So when input comes from the visual external memory system, it is stored in the visual internal memory system. If information is stored in more than one of these systems, memory is improved. Furthermore, memory in the kinesthetic system results in the longest memory (Grinder 1995). Based on this research, Boushey and Moser (2006, 37) developed a set of steps to provide kinesthetic experiences, which are then verbally labeled so students both hear and feel what they are doing, as a way to help students internalize the expectations for each of the components of the Daily Five. As they describe it

> …the class auditorally brainstorms correct behaviors on the I-chart. Then the children model these behaviors in front of the class, permitting them to be seen visually. Finally, the whole class practices these behaviors kinesthetically for three minutes allowing the behaviors to be received and stored kinesthetically for all students through their muscle memories. Daily review of the I-chart, modeling of behaviors and extending practice periods inputs and stores these behaviors in all three memory systems, therefore becoming part of the children's default behaviors.

As teachers encourage student independence when working with mathematics, they can implement the concept of building muscle memory as described above when teaching their students routines and procedures. Students are only successful when they clearly understand the expectations for independent work. The time it takes to teach these expectations is well worth it. Effective steps to teach routines and procedures include the following (Sammons 2012, 229):

1. **Create and post an anchor chart showing what independent work should look and sound like.** Having the class brainstorm what kind of environment they need when working on their own, and how they can support one-on-one conferences as they work, leads them to reflect on the importance of their behavior during these work times. As they share ideas, teachers can record the ideas on an anchor chart and post it in the classroom for future reference.

2. **Introduce and model the routines and procedures with mini lessons.** It is important to link the routines and procedures back to the anchor chart. The links provide the rationales for the routines and procedures teachers establish. It is also imperative that students see the expected behaviors modeled, so they can emulate them as they work.

3. **Have students role-play working independently.** Asking students to demonstrate how they follow the established routines and procedures offers opportunities for them to practice, visualize, and critique the behavioral expectations. Role-playing activities allow students to experience the expected behaviors through all three memory systems—visual, aural, and kinesthetic.

4. **Have students practice and self-assess their behavior.** Teachers can assign independent tasks for students, and then simply observe the students as they work. Initially, work periods of three to five minutes are assigned. Then, the independent work periods are gradually increased. During periods of observation, teachers should refrain from interacting with students, as if they were engaged in conferences and not to be interrupted. At the conclusion of each practice session, students are gathered for a Math Huddle, or class conversation. Students are encouraged to assess their adherence to the routines and procedures and consider how they can improve their work performance.

When students demonstrate the ability to follow the established routines and procedures, teachers can begin conferences with individual students. Whenever necessary, students can be called together to revisit and practice the expectations for independent work that they have learned.

Upholding Expectations for Student Behaviors

No matter how well routines and expectations have been taught, there will be times when they break down. It may be that students become too loud or are off-task as they work independently. Or individual students may interrupt conferences to ask questions of their teachers. For whatever reasons, when established routines and procedures break down, it is impossible for teachers to conduct conferences. When this occurs—and it *will* occur—it is important that teachers not get discouraged and give up on conferring with their students. Instead, teachers should reflect on what is happening in the classroom and look for the root cause of the breakdown.

It may be that the class as a whole needs a refresher on how to behave as they work on their own. Teachers may have to revisit some of the steps they took to teach the routines and procedures. Students can be reminded of the content in the anchor charts that they created or be asked to role-play specific parts of the routines and procedures that are a problem. It may help to re-emphasize the importance of conferences and the role they play in helping both students and teachers. Hoffer (2012, 149) suggests that teacher should give students honest feedback.

> *Share how you feel about being interrupted… Re-explain why conferring is important to you. Discuss again what the class can do to make conferring possible. Talk about the specific incident and brainstorm possible solutions… Affirm that you know the class can do better next time. By keeping it real this way, we communicate to students that we are aware of their behavior and its effects on our classroom culture, that we know they can do better, and that this is what we expect.*

It is important, too, that students always feel the teacher's presence in the classroom. As students begin independent work, many teachers find it wise to spend a few minutes surveying the classroom to ensure that everyone is settling in to his or her assigned tasks (Sammons 2010). If teachers confer where students are working instead of at a table to the side or behind the teacher's desk, learners are more aware of their presence. Sometimes, it helps for teachers to walk around the room for a minute or two between conferences just to remind students that they are there and are aware of what is going on during independent work times (Anderson 2000).

When problems arise from individual students interrupting with questions—a common occurrence, especially during the beginning of the school year—Calkins, Hartman, and White (2005, 31) advise teachers to respond firmly saying, "I was *conferring!*" as if "it is almost inconceivable to me that a child would interrupt something as precious as a … conference."

Teachers are responsible for teaching their students to be self-sufficient learners—not only during conference times. Therefore, students should be learning strategies which they can employ whenever they encounter problems rather than just turning to their teachers for solutions. Sammons (2010, 209) warns, "Without these strategies, students can exhibit a 'learned helplessness.' They immediately ask for help instead of thinking of the options they have and trying to work through the obstacles they believe they face."

Students too often are "spoon-fed" as they engage in mathematical challenges and then lack the confidence and endurance they need when solving problems. So it is imperative that teachers encourage students to assume greater responsibility for their learning. Consistently enforcing established routines and procedures is one way of teaching students to be more confident and proficient learners. In fact, Serravallo and Goldberg (2007, 28) write, "What I try never to do is to engage in a conversation with the [interrupting] student, as this interferes with independence and causes the student to be dependent on me to solve the problem."

When too many students have questions, teachers should re-examine the tasks that have been assigned and check to be sure that students have clear, visual reminders of what they are supposed to be doing as they work independently. Certainly, students will have questions at times as they work. Should questions occur, it helps to have an anchor chart of possible suggestions for students who have nonemergency problems. Possible suggestions for students with questions include:

- Reread the Menu of Instruction (a list of activities or directions for a task) for the math workstation.

- Ask another mathematician in your workstation for help.

- Refer to a math anchor chart for guidance.

- Post the question on the Math Workshop Parking Lot (a chart on which students can post questions using sticky notes to be answered either between conferences or when the teacher has finished conferring with students).

- Choose an alternative task to complete. In your math journal, write the date and tell why you cannot complete the assigned task. (Teachers create a selection of alternative tasks available to students, if needed. Students are expected to record a valid reason why they are unable to complete the assigned task.)

- Ask the classroom technology expert if there is a technology problem. (Teachers may choose to appoint a student who is knowledgeable about technology to assist other students with these problems.)

Chapter Summary

While math conferences are valuable tools for teachers as they strive to maximize students' conceptual understanding, computational fluency, and ability to solve problems, without a supportive classroom environment, teachers find them difficult to manage. It is well worth the time it takes to develop and teach routines and procedures for independent student work. When these are in place, teachers can focus attention on one-on-one mathematical conversations with their students. Students should be taught how to work independently and to handle problems or questions they may have without interrupting conferences.

Because of the brief nature of math conferences, there are many options for including them in a daily schedule. While conferences with students may occur during math periods when students are working independently, either during Math Workshop or engaged in other mathematics tasks, teachers may also choose to confer as students enter the classroom or even during times of transition. Math conferences are an integral component of Guided Math, and the framework offers teachers ample opportunities for one-on-one conferring with learners.

REVIEW AND REFLECT

1. Reflect on your classroom routines and procedures. Do they support one-on-one math conferences? If not, how could you revise them to promote the independent work of students that will allow you to confer?

2. Are one-on-one conferences a tool you presently use for math instruction? If so, what are their benefits? What are the challenges you encounter? What might you do to make it easier to confer with students?

3. Are there any changes you would make in your routines and procedures to help implement or better support Guided Math conferences?

CHAPTER 6

Managing Guided Math Conferences to Promote the Success of Young Mathematicians

The raison d'être of Guided Math conferences is the enhancement of students' mathematical understanding and capability. Earlier chapters have described the many ways in which these math conversations between teachers and students positively impact learning. Yet while the conference itself provides a foundation of insight into student learning and opportunities for teaching next steps in learning, teachers can build upon that foundation to augment the achievement of fledgling mathematicians. The positive effects of the conferences themselves are greatly magnified only when teachers regularly conduct math conferences, maintain an accurate and timely recording system, and then draw upon the data from these conferences to inform their teaching.

Scheduling Math Conferences

Math conferences positively impact students' achievement when they take place on a regular basis. Many of us intend to confer consistently with students, yet find that something always interferes. In spite of our intentions, we may sporadically meet with a student or two—usually only when the needs of particular students demand it. What can we do to make math conferences a priority? One suggestion is to create a schedule. "When we carve out specified times, our plans are a lot more likely to happen" (Slaughter 2009, 82). Dedicating time for conferring and specifically spelling it out in lesson plans can be just the extra motivation needed.

Teachers who are just starting to confer with their students may begin with only one or two conferences a day. Conferences may take a little longer at first than they will later as both teachers and students become more accustomed to conferring. The important thing at this point is simply to make sure that they occur. Teachers' schedules for conferring will vary according to their daily schedules, the frequency with which they choose to confer with their students, and their preferences for when during the day it works best.

Another consideration for teachers as they begin using math conferences is deciding with whom to confer. Who should initiate a math conference—the teacher, students, or both? Most teachers choose to decide with whom they will confer and when. This allows them to be equitable. It is best to confer with all students occasionally, but with some more frequently than with others depending on student needs. What about students who request conferences? Should those requests be granted? Likewise, should students be allowed to defer a conference? Each teacher has to decide how to handle those situations, which will certainly arise. In discussing writing conferences, Anderson (2000, 168) explains, "I felt it was important to give students the opportunity to seek me out, too. Once the school year was under way and I felt that students understood what conferences were about, I invited them to request conferences." Some teachers even create a procedure for students to ask for a conference when they need help and then fit these students into the conferring schedule whenever possible. Whether teachers decide to permit students to request conferences, students should understand that the teacher is never to be interrupted when conferring—even to request a conference or ask for help.

Finally, teachers have to choose when and how they will record conference notes. Teachers who are new to conferring may find it difficult to write notes and hold up their end of the conversation at the same time—especially when conducting research into what students know, can do, and need. Much of these conversations is a spontaneous response to students' comments. While a recordkeeping form helps keep teachers on track as they jot down notes, it may be easier to wait to write notes until after the conference is over. If so, it is wise to remain with the student while recording these notes. With the student nearby, any questions that arise during notetaking can be answered, and interruptions that may postpone or even prevent notes from being written are avoided. Eventually, most teachers find it is best to record notes during the math conference. With practice, they perfect the ability to confer with their students and take notes at the same time.

Keeping Accurate and Timely Records of Conferences

What teachers learn from math conferences with their students is at least as valuable as what their students learn. These focused conversations are "opportunities for teachers to gather the missing pieces of the instructional puzzle" (Sammons 2012, 294). Sadly, what is learned is very often left unrecorded and unused. In an honest statement, one teacher admitted, "I don't take conference notes anymore. I used to write and write and write after conferences, and I never looked at what I wrote again. It just wasn't useful to me" (Slaughter 2009, 85). Without notes recording what was discussed in conferences, the benefits of conferring are limited. Teachers who recognize the valuable role of math conferences in student learning make the effort to develop efficient methods of recordkeeping that work for them and allow them to respond instructionally to what they learn about their students' mathematical thinking.

The Importance of Conference Records

Teachers who have given little thought to how they can use anecdotal conference notes in planning their instruction will have little idea of what kinds of information are of value and should be recorded. Consequently, their notes may be of limited value. It is understandable that these teachers often stop notetaking. Because of this, Calkins, Hartman, and White (2005, 37) advise:

> It is crucial that we don't deflect our attention from the issue of how to keep powerful records by engaging in useless, time-consuming note taking. If we do not have a pressing reason to record a note, we shouldn't write it.

What exactly are the *pressing* reasons for keeping conference records? Clarifying why conference notes are of worth and understanding how to use them is the first step in devising an efficient system of recordkeeping.

Accurate and timely records of math conferences are valuable to educators for the following purposes:

- **To identify needs that should be addressed in later conferences, small-group lessons, or whole-class lessons**

  While teachers may feel as though they will remember what they learned during individual conferences, those who are experienced in conferring have found that not to be the case. Teachers discover that, even with the best intentions, after meeting with several students each day, day after day, these conversations tend to run together. Teachers' days are too packed—with not only their hectic daily schedules, but often with the requirements of teaching several subject areas—for them to clearly retain the details of what was discussed in every math conference. Nevertheless, it is precisely these details that lead teachers to effectively meet student needs.

- **To promote equity**

  In an earlier chapter, the issue of equity versus equality was explored. While equal time when conducting conferences should not be the aim, equity should. It is important for teachers to engage in conferences with *all* of their students—none should be overlooked. But this does not mean that each student must have the same number of conferences with a teacher. It does mean, however, that each student has at least periodic one-on-one conversations about their mathematical thinking with the teacher. Students who are doing well are easily overlooked when teachers confer. Yet these students also benefit from sharing their thinking with their teachers. Conferring with these students reveals the level of their current mathematical proficiency so teachers can plan challenges that are tailored to their specific learning needs.

- **To hold students accountable to previous conference teaching points and learning goals**

  When teachers maintain accurate notes and then refer back to them during conferences, students learn that they are accountable for the teaching points from earlier conferences (Serravallo and Goldberg 2007). Students understand that teachers are looking for evidence that they are employing what was taught during earlier conferences. In addition, because conference records document the self-assessments of students and the learning goals they have established, referring back to them when meeting with learners encourages students to continue to monitor their progress toward their self-set goals.

- **To hold teachers accountable**

  Since conference notes describe the identified needs of students, they are also a means for teachers to hold themselves accountable for responding to those needs. By reviewing these notes, teachers can verify that they have followed up with additional instruction or met with students again to measure their progress toward learning goals.

- **To determine grouping for small-group lessons**

  Teachers who use the Guided Math framework or conduct small-group lessons find conferring invaluable for creating compatible instructional groups. The conference notes document specific learning needs of individual learners. Scanning the notes, teachers can clearly identify students with similar needs. These students are grouped together for small-group lessons targeting those areas. This differentiation of instruction may extend beyond simple reteaching and interventions. Lessons may be differentiated to address varying learning styles, gaps in foundational knowledge and skills, need for additional challenge, and even particular students' interests.

- **To document student growth**

  Since conference conversations reveal student progress, the notes from conferences attest to growth in the mathematical prowess of young learners. This growth is cause to celebrate—with both students and parents. During conferences, teachers may choose to review past notes with students so they clearly realize the progress they have made. Additionally, parents are pleased when teachers recognize and appreciate what their students have achieved. Sharing these positive anecdotal notes often makes parents more receptive if and when student needs are discussed.

- **To document needs and growth for RTI and other intervention programs**

  For students with special needs, it is crucial to monitor learning and assess the effectiveness of intervention programs. Educators providing intervention services need timely, specific, and detailed information about student progress. Information from conferences conducted by either classroom teachers or intervention specialists, when clearly documented, can be shared to ensure that these students are responding well to the services offered and to give service providers data to guide and coordinate their teaching. In the event that students are not progressing, the conference records are valuable resources as teachers work together to design more effective means of intervention.

- **To show students that we care by remembering not only when we have conferred with them, but also what we discussed with them**

  Conferring with students about their mathematical understanding establishes bonds between teachers and their students. As such, it is important that teachers show they care by remembering not only when and how often they have met with young mathematicians, but also recalling what was discussed. When a teacher can say a few words about an idea that a student expressed during an earlier conference or can point out specific growth from one conference to the next, it sends an "I care" message to that student (Slaughter 2009, 86).

- **To provide informal formative assessment of student learning, both individual student and class**

  The benefits of formative assessment are well-documented in educational research (Black and Wiliam 2010; Fisher and Frey 2007; Stiggins 1997, 2002, 2005, 2007). Most teachers realize its importance, but many struggle to find assessment methods that are not overly time-intensive yet provide the kind of data they can readily incorporate into their lessons. Many assessment methods involve too much instructional time for administration and/or require significant teacher time for grading and data analysis. While math conferences should not be the only type of formative assessment used by teachers, these conversations give teachers much deeper insight into students' mathematical capabilities and needs than most types of assessment because of their informal give-and-take nature. Students explain their thinking—often revealing misconceptions and identifying more precisely the instructional next steps that will benefit them most. Since these conferences are brief and notes are recorded immediately, they take minimal amounts of time while providing in-depth information. When teachers review conference notes for their entire class, patterns in learning become obvious that can guide future lessons. For this to occur, however, teachers must maintain an efficient method of keeping conference notes.

- **To share when meeting with parents**

  The notes from math conferences provide clear and specific documentation of the strengths and needs of individual students. Meeting with parents, teachers are able to share anecdotes that evidence their knowledge of each student's learning needs and tell how those needs have been or are being addressed. As Slaughter (2009, 86) writes, "our anecdotal records help us describe the goals and fruits of our work in specific ways." As a result, parents become confident that teachers value their children, are aware of their achievements, and have the professional know-how to meet their learning needs.

Conference Note Forms

Teachers utilize many different kinds of recordkeeping systems for maintaining conference notes. Anderson (2000, 162) states, "From talking with many teachers about their record-keeping systems, I've learned that each teacher needs to use forms that reflect [his or her] own individual needs, tastes, and personalities." It helps to know as much as possible about the systems and forms other teachers use successfully and then borrow the best of them. Using those parts that align with their own ways of working, teachers can create a system that works for them. There is no one right way to keep conference notes. Teachers should bear in mind that, "[w]hat is essential with record keeping is that you keep records and use them" (Slaughter 2009, 87).

Whatever the recording form, it should always show with whom teachers met and what was discussed. Beyond that, teachers should be aware that "any form we carry will influence us" (Calkins, Hartman, and

White 2005, 42). Accordingly, recording forms should reflect a teacher's priorities. As teachers interact with students while taking notes, the focus of the math conference is impacted by what they deem important to record. Recordkeeping forms keep teachers on track and serve as reminders of what they value when conferring with learners.

There are several recordkeeping forms teachers may consider using. Each has advantages and disadvantages. Teachers should feel free to adapt any of these methods of recording conference notes so that they truly reflect their priorities and work well for them. Several of them may be combined or used together. For example, the Guided Math Conference Checklist has little room for notes, but clearly indicates which students have mastered the current mathematics instructional goal. This type of checklist may be used along with another system that has ample space for more detailed notes.

Guided Math Conference Checklist

The Guided Math Conference Checklist (Figure 6.1) offers teachers a way to ensure that no students are overlooked for math conferences (Sammons 2012, 294). Students' names are listed across the top of the checklist. The mathematics instructional goals being targeted are listed in the first column. When a student demonstrates mastery during a conference, the date is placed in the student's column in the row of the corresponding goal. Unmarked cells indicate that students have either not yet mastered the goal or not yet conferred with the teacher. In a glance, teachers can review conference data—helping them determine instructional needs for their class as a whole and for individual learners, as well as identifying the students with whom they have not yet conferred.

With the Guided Math Conference Checklist, the data for the entire class is on just one form, so teachers are unable to gather the notes for each student together in individual student files. This may make sharing data from math conferences with parents more difficult because of the need to maintain the confidentiality of notes regarding other students in the class. In addition, a major drawback to this form is the lack of space for recording more detailed notes about individual conferences. Most teachers use this form to supplement other recording forms that allow conference notes to be more detailed and comprehensive. A full-size version of this form can be found in Appendix C.

Figure 6.1 Guided Math Conference Checklist

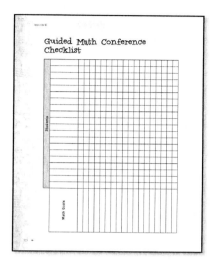

Guided Math Conference Notes

The Guided Math Conference Notes (Figure 6.2) form aligns neatly to the structure of a math conference (Sammons 2012, 294). With this form, teachers record the student's name, the date of the conference, what was learned during the research phase of the conference, the compliment given for an observed strength, and the teaching point. If the focus of the conference was goal setting, the student's goal may be recorded in lieu of a teaching point.

With this type of recording form, it is more difficult to track which students have had math conferences and which have not, but it does provide ample space for documenting what teachers discover during the research phase and what was discussed as a teaching point. As with the Guided Math Conference Checklist, teachers are not able to gather the notes for a single student in one place for future reference or to easily share the notes during parent conferences. A full-size version of this form can be found in Appendix D.

Figure 6.2 Guided Math Conference Notes

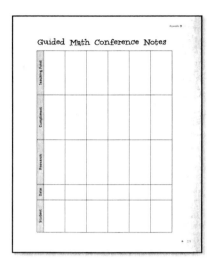

Sticky-Note Organizer

A simple way to keep conference notes is with the use of sticky notes (Sammons 2012, 295). To use sticky notes, teachers fill a sheet of paper with six three-inch square sticky notes spread evenly over the page. These may be placed on a clipboard for jotting down anecdotal notes while conferring. Each sticky note is intended to record notes for one student and should include the student's name and the date of the conference as well as reflect the structure of the conference—including information from the research phase, the compliment, and the teaching point.

Although this system of recording does not clearly show which students have or have not yet participated in math conferences, it does provide plenty of room for notetaking. At the end of the day, teachers can easily scan the notes for patterns of student-learning needs. Which students have similar needs and might be grouped for a small-group lesson? Are there areas in which many students need further work? Should these areas be addressed with a whole-class lesson or review? Or perhaps the notes indicate students are mastering the concepts or skills more quickly than anticipated and could benefit from extra challenges. After reviewing the notes, teachers can place them in individual student files, so that all data for a given student is available in one place. A full-size version of this form can be found in Appendix E.

Figure 6.3 Sticky-Note Organizer

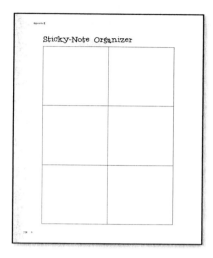

Mailing Labels

Similar to the Sticky-Note Organizer (Figure 6.3), sheets of mailing labels may be attached to a clipboard to record conference notes (Sammons 2012, 295). When mailing labels are used, teachers have the option of preprinting them with student names—one per student. By checking the empty labels, teacher can see clearly the students with whom they have not yet conferred. In addition to preprinting these labels with student names, the labels may also be printed to include the following categories: date, research, compliment, and teaching point. See Figure 6.4 on the following page for an example of how mailing labels can be used to keep notes during Guided Math conferences.

Figure 6.4 Sample Mailing Labels

Name *Maurice* Date *January 10* Research *Tells time accurately to the half hour* *Has quarter hour —15, but not 45* Compliment *Identifying the hour and minute hands, using* *hour hand to determine hour* Teaching Point *Reviewed that 5 minutes between each* *number, practicing skip counting by fives to* *determine minutes after the hour*	Name Date Research Compliment Teaching Point
Name Date Research Compliment Teaching Point	Name Date Research Compliment Teaching Point
Name Date Research Compliment Teaching Point	Name Date Research Compliment Teaching Point

Sectioned Notebook

Another practical recording system is the use of a thick spiral notebook. Tabs for each student are attached every few pages creating sections of about five pages for each student. When teachers confer with a student, they simply open to that student's section, note the date, and add whatever information from the conference they think is important. This is an easy method to manage and store the notes for individual students together, making it convenient to access. On the other hand, relying on this method of recording makes it harder to identify students who might have been overlooked for conferences or to scan the daily conference notes to identify learning patterns. See Figure 6.5 for an example of a Sectioned Notebook.

Figure 6.5 Sample Sectioned Notebook

Student Flipchart

Some teachers choose to keep conference notes on a clipboard flipchart. These flipcharts may be created by taping slightly overlapping five-by-eight-inch index cards onto a clipboard (see Figure 6.6). One card is taped at the bottom of the clipboard. Other cards are added, each one taped about a quarter of an inch above the previous one. This leaves visible the lower edge of each card upon which student names or numbers are written. Some teachers choose to list their students in alphabetical order and then assign numbers to them beginning with one. Students are asked to place their names and numbers on their work to be turned in, making it easier for teachers to put student papers in alphabetical order. Using a number system also allows teachers to reuse files and other materials labeled in this way from year to year. If teachers choose to assign student numbers, each card may be labeled with a number instead of a name. There should be one card for each student in the class. To record conference notes, teachers just turn to the card of the student with whom they are conferring and jot down their notes. When a student's card is filled, it is removed and placed in the student's file. And then, a new card is taped to the clipboard to replace the one that was removed.

Although this method does not allow teachers to monitor which students have had conferences or to easily scan the conference notes for the day, it does make the notes more private. The only note visible to students who might be looking over the teacher's shoulder is the one that is being written. Students tend to be curious about what their teachers are writing, and many do not hesitate to read the conference notes for other students if they are in view. While this may not be an issue for young students, for older students who are better readers, it is something for teachers to be aware of.

Figure 6.6 Sample Student Flipchart

Electronic Notes

With the technology available today, more and more teachers are turning to electronic devices, particularly electronic tablets, for notetaking. Programs such as Evernote® and Noteshelf offer teachers sophisticated ways to record conference notes, some of which offer password protection to maintain the confidentiality of student information. Folders may be created for each student in the class where their conference notes are filed. Notetaking forms can be designed, saved as templates, and then easily pulled up for use. With some of these programs, notes may even be handwritten using a finger or stylus.

If teachers wish to save student work samples, they can take pictures and include them in the electronic file. Short videos of students discussing their work may also become a part of conference records. Many of these programs contain a search function allowing teachers to easily search notes for specific words or phrases. Using this function, teachers can locate and group students with similar needs for small-group lessons. An additional advantage of using electronic notes is ecological—no paper is used.

Refining Conference Notetaking Skills

Once teachers have decided on a recordkeeping system that works well for them, the next step is to hone their notetaking skills. As mentioned previously, conference notes tend to reflect teachers' priorities. Although it requires a little extra time, it is well worth the effort to examine the conference notes taken periodically and consider what is being recorded and what may be missing from these notes. Do the notes truly reflect the teachers' priorities *and* the learning goals for the student? Do the notes tend to get bogged down in procedural detail rather than document the students' understanding of mathematical big ideas or their ability to apply those ideas in other contexts? Do the notes give information that will help determine where on the learning progression toward grade-level learning goals the student falls? Are the notes specific enough that several strengths and needs are clearly identified? Do the notes provide guideposts for future instruction? This reflection can be an interesting exercise for teachers—somewhat "like watching our own thinking" (Calkins, Hartman, and White 2005, 42).

Because teachers' conference notes so clearly reveal their own thinking, it is worthwhile for them to meet with their peers to examine conference records together. Discovering the diversity in what other teachers choose to note offers new insights into the information that might be of value to record. Challenging teachers to expand their vision as they take writing conference notes, Calkins, Hartman, and White (2005, 43) question, "[M]ight you deliberately want to try to alter your governing gaze so that your teaching takes into account new aspects of the child, writing development, and of your teaching?" Just as with writing conferences, teachers may want to explore new avenues of mathematical development. Since conference notes directly impact future lessons, new perspectives in notetaking inevitably lead to new perspectives in teaching mathematics as well.

Using Math Conference Notes to Plan Instruction

One of the main reasons teachers choose to confer with their students is so they can pinpoint and target students' mathematical needs more precisely. For differentiation of instruction based on information from conferring to be most effective, however, creating and maintaining quality conference notes is essential. Only when care is taken to record meaningful details from their mathematical conversations with learners can teachers use what they learn to its fullest potential. Part of this benefit arises from the recording process itself as teachers reflect and weigh the importance of what was discussed. Later, these notes afford valuable opportunities to review what they know about their students, to consider how to address their students' needs, and to fine-tune their mathematics instruction.

Hattie's (2012b, 25) research shows that expert teachers, "make lessons uniquely their own by changing, combining, and adding to the lessons according to their students' needs and their own teaching goals." Whether we are planning individual conferences, small-group lessons, or whole-class lessons, when wisely used, conference notes allow us to "see learning through the eyes of students" and plan accordingly (111). The key phrase in that sentence is *wisely used*. Anderson (2000, 160) reminds teachers, "The notes I take are only useful, of course, if I actually refer back to them." Only when conference records are well organized, readily accessible, and, yes, *used*, does this instructional tool have the greatest impact on learning.

Planning for Individual Students

Because each student's needs are unique, only when we really know our students can we most fully meet their learning needs (Sammons 2012). Upon reviewing conference notes, teachers may choose to plan for individual follow-up in any of these ways:

- Recheck conferences may be conducted with some students to ensure that the teaching point of a conference was understood and being used. If not, the teacher can provide additional support for the student as a teaching point during this conference. If so, the teacher may decide on a teaching point that will extend the student's understanding or skill related to the previous teaching point.

- During some conferences, teachers may notice several areas that could be possible teaching points, but could only address one at that time. When this occurs, teachers may choose to confer with the student again to focus on the other teaching points.

- When conferences show that students need additional scaffolding to access the math content, teachers may provide it either during another conference or with one-on-one instruction. It is important to note that anytime teachers provide scaffolding to support student learning, they should also carefully plan how it will be withdrawn. Scaffolding is only a *temporary* support for students. The goal of scaffolding is to assist students in their progress toward proficiency when working independently. To achieve this, there must be gradual, incremental, and well-planned withdrawal of the scaffolding rather than an abrupt removal of support. Providing plans for the gradual withdrawal of support frequently involves planning steps to move students from a concrete, to a representational, and then, finally, to an abstract level of understanding.

- Conferences may reveal several areas of concern regarding a student's math progress. When this happens, teachers should be attentive to those aspects of the student's work throughout the school day to monitor his or her understanding and skill. To further support the student's mathematical growth, the teacher can involve the student in tasks that will reinforce his or her ability in those areas of concern.

- If students set math goals for themselves during goal-setting conferences, teachers should continue to monitor their progress in meeting their goals during their ongoing mathematical work when working independently. When students achieve self-established goals, celebrations are in order. The entire class, as a mathematical community, can be encouraged to join in the recognition of every student's accomplishments. Shared celebrations provide additional impetus for other students who are still working to attain learning goals.

- Math conferences may reveal a student's burning interest in a math-related area. If so, teachers should feed the fire by connecting grade-level math topics to the student's area of interest (Sammons 2012). Students should be encouraged to explore areas of particular interest. One student's strong mathematical interest may well stoke the interest of others.

Planning for Small Groups

When teachers confer one-on-one with their students, the insights gained should be used to flexibly group learners for targeted instruction with lessons tailored to specific learning needs. Some of the many ways in which math-conference data serves to support small-group lessons are the following:

- If data from testing indicates a number of students missed certain questions, yet fails to show why, math conferences with these students permit teachers to investigate the reasons behind the errors. As students explain their justification for the erroneous answer choices, teachers glean the root causes of their errors: gaps in background knowledge or skills, misconceptions, careless mistakes, or lack of computational proficiency. Sometimes, errors in solving story problems even stem from difficulties reading and interpreting the problem. If certain error patterns are common for several of these students, teachers may choose to meet with them for a small-group lesson focused on eliminating the identified causes of the errors. Without having conducted conferences prior to the small-group lesson, teachers may either use small-group lessons to search for the causes of the errors or else reteach based on guess work—not knowing what the true instructional needs of these students are. But once the causes of the errors are determined by meeting one-on-one with students, they can be targeted efficiently in small-group lessons "rather than [using] the valuable time with your small group to probe for underlying causes of errors" (Sammons 2012, 299).

- When math conferences reveal several students who have quickly understood the content being taught and are ready for greater challenge, teachers can group these students for small-group lessons to lead them to think more deeply and to accelerate their learning. The needs of these students are too often unaddressed because of teachers' focus on providing interventions for struggling students, and because without math conference conversations, their need may not be apparent.

- Math conferences conducted prior to the introduction of a new concept or skill may indicate whether some students have gaps in the prerequisite background knowledge or skills they need to be successful with the upcoming content. Grouping students with shared gaps enables teachers to target and fill them before teaching the new mathematical concepts or skills.

- Math conferences sometimes reveal the unique, and often compelling, mathematical interests of students. When clusters of related student interests are evident, these learners can be grouped together for small-group lessons based on these mathematical areas of interest.

- The learning styles of students differ considerably. While students need experiences honing diverse learning modalities, it may be helpful, at times, to provide occasional lessons for small groups of students to support their predominate learning styles.

Planning for the Whole Class

Although math conferences are between just one student and the teacher, the data gathered from conferring one-on-one with many students offers teachers a fairly comprehensive overview of class needs. Some of these needs may not be obvious from the feedback teachers obtain through other modes of instruction or through more traditional assessments. Teachers maximize the benefits of math conferences when they step back and look at the big picture—look beyond the trees to the forest.

Math conferences may reveal that quite a few students in the class are struggling with the same concept or skill. If so, teachers may want to confer even further with more students as a pulse check of the class on that particular concept or skill. Are other students experiencing the same problem? When the struggle appears to be widespread, teachers may choose to provide extra support or reteaching for the whole class. If conference notes show that some students have no need for additional support, other more appropriate mathematical tasks may be provided for them.

Similarly, if a teacher's conferences with students show that many of them demonstrate an understanding beyond what was expected, it may be worthwhile for teachers to confer with additional students to get a feel for whether this is true for most of the class. Depending on the findings from these conferences, the teacher may decide to move ahead with instruction a little more quickly than originally planned or to increase the level of challenge to respond to students' degree of understanding and skill. Effective teachers use textbook resources flexibly. When these resources call for additional lessons on certain content, but student mastery of it is evident from the math conferences conducted, teachers should use their professional judgment about whether each and every lesson is taught (Sammons 2012).

One of the more interesting aspects of conferring with students is learning about their individual impressions of the math concepts with which they are working and their often unique tactics for problem solving. Teachers should be open to considering student perspectives that may be valid albeit untraditional. It often generates student interest when teachers share these perspectives and tactics with the whole class. Even when lacking in validity, the discussions generated as students attempt to justify their reasoning provide fodder for rich whole-class lessons. Viewing a math concept or procedure through the student's lens may suggest lessons that approach the concept from a different angle or may suggest "I wonder…" questions (Sammons 2012). The teachable moments that arise in math conferences may be parlayed into engaging whole-class lessons, making them both more relevant and motivating to students.

Chapter Summary

The value of conferring with students about their mathematical thinking extends far beyond the conference conversations themselves. The information accrued when conferences are conducted on a regular basis can be used to ensure mathematics instruction effectively meets the needs of students. A system to ensure accurate and timely recordkeeping is essential if this is to occur. Teachers need to consider what information from conferences is important to record, and then maintain conference records in such a way that they may be easily accessed and used for planning instruction. By reviewing well-maintained conference notes, teachers are able to identify not only the needs of individual students, but also of clusters of students, and even of the class as a whole.

Ascertaining varying instructional needs is only the first part of the process. Without using this assessment data for planning instruction, it has little value in supporting the mathematical learning process. When teachers pinpoint instructional needs—whether they involve filling gaps, providing additional challenge, sharing interesting approaches to understanding concepts and problem solving, responding to student interests and learning styles, or revisiting mathematical concepts and skills, their next step is deciding how to most successfully address these needs. Is the need limited to a single student or to a small group of students? Or is the need more prevalent in the class as a whole? Answering these questions directs teachers as they decide how to respond to the needs of their students. And finally, guided by what they learn from math conferences, teachers design their lessons with the conference data in mind.

REVIEW AND REFLECT

1. Do you conduct reading or writing conferences with your students? If so, how does it impact your instruction? What kind of recordkeeping system do you use? How effective is it? How might this be adapted to maintain math conference notes?

2. In deciding whether you will conduct math conferences regularly, you may choose to do a trial run. Choose three students randomly from your class with whom you will confer. Conduct the conferences and carefully record the research findings, compliment, and teaching point from each. Examine the information you obtain and decide how you might use it to tailor your instruction. When you complete the trial run, reflect on the experience. What have you learned from it? How did it impact the students with whom you conferred? How might it affect your teaching?

CHAPTER 7

Guidelines for Effective Guided Math Conferences

Teachers who choose to confer with their students because of its documented benefits continuously work to make their conferences more engaging, more beneficial for their students, and more revealing as a formative assessment. This is not an easy task. Student personalities and learning needs are diverse. As Calkins, Hartman, and White (2005, 6) recognize, "[c]onferring is always a challenge."

With the most successful conferring, teachers maintain an awareness of the learning goals for their students, let students know that they genuinely care about them and are interested in hearing their mathematical ideas, ask carefully-crafted questions that prompt students to think more deeply, encourage students to monitor their own learning and set goals for themselves, as well as deciding on teaching points and sharing them with students. Whew! What an undertaking!

But consider the process of walking. If broken down into each essential component, it seems impossible to do—the contractions of muscles, the balance needed as weight is shifted from one foot to the other. How are people able to put all of these together to walk? Yet walking is easily achieved by most of us. Similarly, teachers should be reassured—the actual process of conferring is not as difficult as it seems. Teachers often find that their instincts as teachers take over during conferences. Not that they go into conferences without having in mind some areas of focus, but by drawing upon their professional expertise, responding flexibly, and listening to the mathematical conversations of students, their conferences are very effective. Nevertheless, there are some guidelines for successful math conferences.

Figure 7.1 Guidelines for Effective Guided Math Conferences

- Recognize the difference between helping and conferring
- Communicate as a fellow mathematician
- Make conferring predictable
- Listen actively to young mathematicians
- Ask questions that reveal and extend thinking
- Encourage students to do most of the talking
- Strive to understand the thinking of students and follow their line of thinking
- Build on student strengths and interests
- Teach just one thing
- Avoid power struggles
- Set clear expectations for students
- Encourage deeper and more complex thinking
- Celebrate mathematical growth
- Incorporate humor and playfulness
- Use conferences strategically

Recognize the Difference between Helping and Conferring

Teachers preparing to confer with students or hoping to improve their math conferences should consider the distinctions between helping and conferring. Hoffer (2012, 143) warns that many teachers who engage in one-on-one conversations during math lessons are "trapped in the typical pattern of helping that leaves students dependent and awaiting our approval." With conferences, teachers spur students to think and communicate about their mathematical thinking by listening respectfully and asking open-ended, thought-provoking questions. Teachers converse with their students as one mathematician to another with the aim of expanding and deepening their students' mathematical knowledge. Rather than students viewing teachers as the repositories of mathematical knowledge upon which they are dependent, teacher-student math conferences ideally lead students to see themselves as capable mathematicians who can draw upon what they know to solve problems. When teachers clearly discern the difference between just helping students find correct answers and conferring to enhance mathematical learning, then conferences are at their best. Figure 7.2 offers a comparison between helping and conferring.

Figure 7.2 Differences between Helping and Conferring

	Helping	**Conferring**
Focus	Getting answers	• Teacher learning about student thinking • Student communicating, learning, and reflecting to deepen his or her mathematical understanding and skill
Expectations for the Student	• Listening to the teacher • Asking questions (if needed)	• Sharing his or her thinking • Pondering current and new ideas • Making mathematical connections • Trying new strategies • Extending his or her mathematical understanding and skill
Expectations for the Teacher	• Telling or showing students how to improve their work • Pointing out errors • Reteaching	• Active listening • Prompting deeper thinking by students with questions • Sharing feedback with students • Identifying and teaching a next step in learning
Goal	Students get the correct answers.	• Students thinking more deeply and learning mathematics • Teacher gaining deeper understanding of students' mathematical capabilities
Inferred Beliefs	Teachers are the dispensers of mathematical knowledge and how-to.	Students are mathematicians with interesting ideas to share and the capacity to solve problems.

(adapted from Hoffer 2012)

Communicate as a Fellow Mathematician

Imagine two educators engaged in a conversation about teaching. Most likely, these colleagues are seated close to each other. They talk amiably about what works well for them in their classrooms, about the challenges they face, and about what new teaching ideas they may try with their students. They show their respect for each other by listening closely and asking questions to clarify their understanding when needed. They may ask for each other's opinion about instructional strategies. While either teacher may make suggestions, neither teacher assumes to have all the answers to the challenges they face. Both teachers, however, know that they can learn from each other.

The kind of easy amiability evident in the imagined conversation just described is what teachers should strive for when conferring with their students. Ideally, Guided Math conferences are conversations between fellow mathematicians talking about mathematics and their understanding of it. Their talk should reflect a true interest in and respect for each other's mathematical ideas. When trusting conference relationships exist between teachers and students, teachers are able to probe student thinking in a nonthreatening way. Students feel comfortable sharing untested ideas or admitting a lack of understanding. And teachers are able to do something that is often very difficult—they can let their students know that sometimes they do not have the answer to a mathematical question. In fact, they may choose to use these occasions as teachable moments, modeling how one searches for the answers to questions.

One way to establish this kind of a relationship is by paying close attention to the conference setting and climate. According to Graves (2003, 97), "[t]he conference setting is a form of language." It sends clear messages to students. Simple steps to establish a conference setting conducive to building trusting relationships include:

- **Confer at the desk or table where the student is working.** Not only does this avoid students being called to meet in what they might consider "teacher's territory," but it also creates minimal disruption to their work. "When your students feel like your conversation is a part of the work they were already doing before you arrived … they can easily and smoothly return to independence when you move on" (Hoffer 2012, 144). Furthermore, according to Anderson (2000, 156), "students seem more at ease talking with me when I meet them where they feel most comfortable, at their desks or tables, even on the floor in a corner." Conferring with students where they are working has another added advantage for teachers. Serravallo and Goldberg (2007, 29) write, "positioning myself physically in the room allows me to manage my class a little better. My presence is felt."

- **Sit side-by-side with the student.** With this seating arrangement, both parties to the conference can easily look over the mathematical work being discussed. Sitting across from a student, rather than beside, may send a message that the teacher does not wish to be closer or that the teacher is in an adversarial position (Graves 2003). Sitting beside students for conferences rather than standing above them also defines a teachers' relationships with their students—putting them in the position of talking *with* their students, not *down* to them (Anderson 2000).

- **Speak in a "conference" voice.** Sitting at students' eye level puts students at ease. Because of their close proximity to students, teachers can speak softly so their voices are no louder than those with whom they are conferring. It may be intimidating to students if teachers speak with the powerful voice they use for whole-class instruction. On the other hand, a softer and gentler voice shows students that teachers respect them and are interested in hearing about their work (Anderson 2000).

- **Ask the student to tell about his or her work.** Beginning the conference by asking a student to explain his or her work sends the message that the student is responsible for his or her work. Rather than picking up a student's work to examine it, it is more respectful to leave it in the student's hands and to encourage him or her to assume responsibility for it.

- **Acknowledge and respond to mathematical ideas and concerns the student introduces.** Teachers should temper their own instructional agendas when they are conferring to acknowledge the ideas and concerns that students express. By their very nature, conferences are *two-way* conversations. Although teachers should have in mind what they would like to accomplish as they confer, by being open to student interests and concerns, they establish stronger relationships with their students and learn much more about their learners' mathematical needs.

Make Conferring Predictable

Students, like so many others, are most comfortable when they know what to expect. Making conferences predictable—when and where they occur, how they begin, and how they are structured—ensures that students are prepared for them. No time is wasted as learners try to figure out what is going on or what is expected of them. Math talk tends to flow more freely because both teachers and students can predict the structure, although not necessarily the content, of the conference (Anderson 2000).

Teachers benefit from a consistent conference structure, too, because it means that the same conversational approach has been followed in their conferences with many students (Graves 2003). Students' thoughts shared during these conferences can then be viewed in light of those of other students, allowing teachers to more accurately determine whether the learning needs they identify are those of an individual student only, of a small group of students, or of the class as a whole. The predictability of conference structure also makes it easier to monitor the progress of individual students as their conversations can be more easily compared.

Listen Actively to Young Mathematicians

With conferring, the tables are turned for both students and teachers. Traditional instruction demands that students listen carefully as teachers lecture and only respond when asked. With math conferences, however, teachers must listen intently as their students talk about math. Changing roles may be harder for us to do than one would expect. Fletcher and Portalupi (2001, 49) point out that "about 75 percent of what we do as teachers has to do with what was done to us at the other side of the desk, when we were students. And few of us had teachers who truly listened to us when we were kids." To be successful with this reversal of roles, most teachers have to practice in order to hone their listening skills.

Listening to children is more a deliberate act than a natural one. It isn't easy to put aside personal preferences, anxieties about helping more children, or the glaring mechanical errors that stare from the page. I mumble to myself, "Shut up, listen, and learn!" (Graves 2003, 100)

When conferring with students, we may feel rushed, see multiple student errors or misconceptions that need correction, and very often just want to get to the teaching point. While we really do want to be good listeners, the responsibility of monitoring the rest of the class also distracts us. Anyone who has ever taught, or even multi-tasked, understands the struggle to be a good listener. Active listening, especially in a classroom setting, requires both intent and practice.

Active listeners "not only hear the words that another person is saying, but more importantly, try to understand the complete message being sent" (Mind Tools, under "About Active Listening"). There are five specific things we can do to increase our understanding of what a student is sharing and to convey to the student that they are being heard (Mind Tools). Teachers are listening actively when they:

- **Pay close attention to students with whom they are conferring.** It may be difficult with the commotion of a typical classroom, but when conferring, teachers should give students their undivided attention. Looking directly at students while they speak lets them know that they have the teacher's attention. While it is tempting to be distracted by elements of what students are sharing, it is more beneficial to attend to the totality of their ideas before trying to address specifics.

- **Show students that they are listening.** A teacher's body language and gestures can convey a listening attitude. As students are speaking, teachers should nod occasionally, smile warmly, maintain a posture that is open and inviting, and encourage students with brief verbal responses.

- **Give students verbal feedback.** When teachers reflect and briefly summarize what students are sharing, it not only shows students they are being heard, but it also allows them to clarify any misunderstandings teachers may have about what has been said. This is especially important because many students lack the mathematical communication skills to clearly express their thinking.

- **Let students finish their thoughts before asking extensive questions or beginning to share a teaching point.** Many times, questions that teachers have initially are answered if they wait and listen. Teachers demonstrate respect for their students by waiting patiently for them to complete their thoughts. Although in a fast-paced classroom atmosphere it may be difficult, much is to be gained by slowing the pace for conferring. When students feel rushed, they are less likely to make the effort to compose their thoughts and share them during their math conferences.

- **Respond respectfully to students' statements.** Being candid with students about their mathematical ideas is important, yet it must be done with sensitivity for the conferees. Students may be trying out new ideas or strategies—some of which may be successful and some of which may not—and taking risks. For this, they should be lauded. In many cases, well-considered questions by teachers effectively lead students to discover their own errors, reducing the need for teachers to point them out and then correct them.

- **Focus on understanding students' thinking, not correcting it.** There will be many times when it is difficult to understand the ideas students are expressing. These may be awkward for both teacher and student. It is important to resist the temptation to ignore the confusion and move on. "Unless you persist in trying to get clarification, students will not learn that you intend to take their explanations seriously" (Chapin, O'Connor, and Anderson 2003, 124).

Developing these listening skills requires practice. But their value when conferring with students is great. If we truly want to learn about our students' mathematical knowledge and skills and, at the same time, encourage them to think about mathematics with more depth and complexity, it is well-worth the effort to practice active listening when conferring with learners.

Ask Questions that Reveal and Extend Student Thinking

Questioning shapes conferences—and, consequently, determines what both teachers and students learn. The questions themselves are not the goal, but only the by-products of teachers' efforts to learn more about their students' mathematical thinking. "When questions grow out of our emerging understanding of the writer, they are alive and fresh and powerful" (Calkins 2000, 225). Similarly, math conference questions should grow from an understanding of the mathematician. If questions are randomly selected from a list, they "quickly become canned and mechanical" (225).

Many of us had poor experiences with questioning when we were in school. Our teachers asked questions that had only one correct answer. We knew that. Those of us who thought they knew the answer were eager to respond; those who were not sure sought ways to avoid answering. Obviously, that kind of questioning is not conducive to productive conferences. Instead, as Graves (2003, 107) notes, "[g]ood questions provide surprises for both child and teacher. The child finds himself speaking about information he hardly knew he possessed. The teacher may have had only an inkling that the child knew the information." Questions that reveal and extend student thinking are timely and often spontaneous, coming in response to students' statements and teachers' desires to know more about their students as mathematicians.

Open-ended questions guide students to an understanding of the complexities of mathematics and how what they are thinking and doing mathematically fits into the bigger picture. When teachers ask questions such as, *What are the big math ideas you are thinking about as you work today?* or *How are they connected with what you are doing?* teachers are letting students know that they see them as mathematicians whose ideas have value.

In general, there are eight kinds of questions that teachers ask during math conferences (Graves 2003).

Opening Questions

These questions are typically consistent and predictable. Teachers decide on an open-ended question they use to begin their conferences. Students know what to expect and are comfortable responding. There are times, however, when teachers opt to begin with different questions based on what they know about the student with whom they are speaking and the specifics of the conferring situation. Students who lack self-confidence may not be able to respond well to the same questions other students take in stride. Teachers should exercise their discretion and deviate from their standard opening questions if it seems appropriate. These questions work well to begin conferences:

- *How is your math work going?*
- *What are you working on?*
- *Will you share what you are thinking about mathematically today?*
- *You seem to be (frustrated/stumped/at a standstill). Can you tell me why?*
- *I noticed…. Will you tell me about it?*

Following Questions

Some students talk on and on about their math work with little need for encouragement. Other students depend on questions to keep them going. Teachers may use the following questions to "nudge students to say more" (Anderson 2000, 42). These are most effective when asked spontaneously in response to students' comments. Composing these questions is easier for teachers, however, if they already have in mind some queries that might serve this purpose. Questions like these prompt students to add more to the conference conversations:

- *Can you tell me a little more about…?*
- *What made you decide to…?*
- *Why do you think…?*
- *If you could use only one word, how would you describe your math work? Why did you choose that word?*

Process Questions

These questions help students become more conscious of how they function at mathematicians. They prompt students to explain the process of thinking mathematically—not in the abstract, but when viewed through the unique prism of their own personal mathematical experiences (Graves 2003). Teachers may decide to ask one of these questions when a student has just solved a problem. By describing the processes they used, students are honing their mathematical communication skills. Their reflections as they consider how to describe what they have done serves to reinforce their mathematical understanding. These questions are examples of process questions:

- *What strategy did you use to solve this problem? Why did you choose it?*

- *I see you tried several different approaches to solve this problem. Which one worked best? Why?*

- *How did you know that…?*

Questions that Reveal Development

Sometimes teachers ask students directly to assess their growth as mathematicians. At other times, probing questions will lead students to discover just how far they have progressed. These kinds of questions are effective only when students are aware of their learning goals. Highlighting students' mathematical development through questioning is beneficial to both students and teachers. Students gain confidence when they are meeting learning goals. Teachers gain a greater understanding of their students. If students are struggling to meet their learning goals, teachers should encourage them to make a plan for achieving them.

This kind of questioning serves as formative assessment—allowing teachers to monitor learning so their lessons best meet the instructional needs of their students. Questions like these help reveal students' mathematical development:

- *Think of your learning goals. How do you think you are doing toward meeting them? What do you feel most sure of? Is there anything that is giving you trouble?*

- *Do you remember when we began this unit? What did you know about _____ then? What do you know now? What have you learned?*

- *How does what we are learning about _____ help you with the math you are doing today?*

Questions that Address Knowledge of Mathematical Concepts and Skills

Teachers may have to nudge students to get them to explain what they know about the mathematics with which they are working. Most of the time, this information can be determined by listening closely to the explanations of their reasoning. At times, however, it is not obvious whether young mathematicians clearly understand the concepts with which they are dealing or know how to apply those concepts and skills to the tasks at hand. Questions that address mathematical concepts and skills are used to find out. Intended to shed light on students' basic mathematical knowledge, they are more specific and less open-ended than most conference questions. Questions like those listed below address learners' knowledge of the basic structures of mathematics.

- *What exactly is (are)…?*
- *What does _____ mean? How does it apply to what you are doing?*
- *What do you think are the most important aspects of these mathematical concepts?*
- *What steps did you take when you…?*
- *What operation are you considering using? Why?*
- *You mentioned _____. In what ways does that relate to your work?*
- *You just _____. What will do you do next?*

Questions that Extend Thinking

If students seem to be floundering or reaching a dead end with their work, the use of strategic questioning may help them recognize patterns, make connections, draw upon mathematics comprehension strategies, or apply what they already know and can do to move ahead with their work. These questions can also prompt students, especially those who may need additional challenges, to think more deeply about mathematical concepts with which they are working (Sammons 2012). These questions are often asked as part of the teaching point of a math conference.

- *How is this math similar to other mathematics you have learned about?*
- *What questions could you ask that would help you understand it better?*

- *What do you think are the most important aspects of this problem?*

- *Do you notice any patterns?*

- *Can you make a prediction based on what you know?*

- *How else can you represent your mathematical thinking?*

- *How do you visualize the problem you are working on?*

- *As a result of your work, do you have any new mathematical ideas?*

Questions that Pose a Challenge

Graves (2003, 116) reminds writing teachers, "Not all problems are solved in conference. Sometimes it is important to cause problems, problems that are solved *outside of the conference.*" Rich opportunities for genuine mathematical investigations may arise from one-on-one math conversations. Building on ideas discussed, teachers can pique the curiosity of their conferees with mathematical questions that are not easily answered. Asking students to search for answers after the conference encourages their self-sufficiency. When mathematical challenges move beyond the conference confine, it is amazing how student interest in them becomes contagious. Teachers can make the most of this by encouraging students involved in these mathematical quests to share what they are doing with the class. Although questions that pose challenges most often flow directly from the subject matter of conference discussions, these scenarios give some ideas for the kinds of challenges that teachers may pose to students:

- *You noticed a pattern. Can you make a conjecture based on that pattern? Do you think it will always be true? For the next few days, see if you can find any instances when it is not true.*

- *The problem you just solved involved percents. During our conference, you mentioned that you don't see why you have to learn how to work with percents. Over the next few days, how many times do you think you will see percents (outside of the math work you are assigned)? Why don't you keep a list for a couple of days? I am curious about what you find. When you finish, would you mind sharing it with the class?*

- *As we talked today, you said that at first you thought your answer was wrong when you multiplied two times one-half and found that the product was one. You explained that you had learned that whenever you multiply, the product will be greater than the factors. Will you keep thinking about that to see if you can come up with a clear reason as to why the product was less than one of the factors for this multiplication problem?*

- *You shared a lot of ideas with me during our conference today about making patterns using three different colors. Do you think you have thought of all the possible patterns you can make? Why don't you keep on working on it? You can talk about it with other kids if you want. Whenever you get a chance, keep trying to find more patterns. Be sure you make drawings of them or represent them in some way so you can show me what you find.*

Questions that Encourage Reflection

One of the goals of math conferences is teaching students how to think about both mathematics and their understanding of it. The focus of all math conference conversations should be on student thinking. Teacher questions can help students clarify their thoughts. Particularly during the linking phase of conferences, students should be asked to reflect on what they learned during the conference and how they will use it in the future. Reflection questions may address either understanding of the discipline or students' metacognition.

- *Does your solution to the problem make sense? Tell me why you think so. If not, what will you do?*

- *How well do you understand what you are working on?*

- *Do you feel the math you are doing is challenging enough? Why or why not?*

- *What about the mathematics we are studying is exciting to you?*

- *What did you learn or think about mathematically during our conference?*

- *How will you use what you have learned in the future?*

- *What do you wonder about regarding the mathematics we are studying?*

- *What are your math goals?*

Encourage Students to Do Most of the Talking

One way of making math conferences more productive is by encouraging students to do most of the talking. This gives young mathematicians valuable experience organizing and then communicating their thinking. Teachers who signal their expectation that students have important ideas to share and then listen to them attentively demonstrate respect for students as mathematicians and establish supportive relationships. Additionally, because student talk is such a large part of the conference, teachers learn much about students' mathematical proficiencies and needs—making conferences one of the best methods of formative assessment.

Many students have grown accustomed to the expectation that their role in school is to listen to teachers and respond only when asked specific questions. Graves (2003) describes a process of *reversibility* in which this assumption by students may be changed if teachers simply wait for their students to respond to open-ended questions. Sometimes, teachers may have to restate their questions, but giving students ample wait time clearly signals to students that teachers really want to hear about their mathematical thinking. When teachers wait for their students to respond, the responsibility for explaining their thinking shifts to learners. They learn to initiate mathematical conversations during conferences with little or no prompting and become comfortable asking questions they have about their mathematical work.

Thus, for effective conferring, teachers must "be prepared to get used to silence" (Graves 2003, 99). This silence may be uncomfortable and even unnerving for teachers. Considering that students are usually given three to five seconds for a response to a teacher's question, lengthening that considerably (to up to forty seconds or more) takes patience on the part of teachers. With this extra time, however, students have an opportunity to formulate their answers. When students know that teachers are going to wait for their responses, it becomes part of the predictable nature of conferences. According to Graves (2003, 99), "[c]hildren will use silence when the conference is predictable, when the setting is right, and when they believe you think they have something worthwhile to say. Just wait for it."

Providing students with predictability in regard to the questions teachers ask also encourages students to do the talking. Often, it's the repetition of familiar questions that cues students to share their mathematical thinking, not necessarily the questions themselves. Using predictable questions takes advantage of "students' implicit knowledge of the nature of conversations" (Anderson 2000, 29). Along with predictability, there are other conversational strategies that teachers can employ to support students' mathematical conversations. Figure 7.3 presents an overview of these strategies.

Figure 7.3 Conversational Strategies to Support Mathematical Conversations

Conversational Strategy	Who It Supports	Examples of Teacher Support
Redirect	Students who shift the conversation away from mathematics or their own understanding of mathematics	"I bet you had a great weekend, but tell me more about what you are working on." "How does what Max did to solve the problem compare to what you are doing?"
Reflect and Pause	Students who have some facility with talking about their mathematical work	"So your math work is going well…" "I see…"
Show and Describe		"Will you show me what you are working on today, and as you do, describe what you are doing?"
Refer Back to the Last Conference		"When we last talked, you were using base-ten blocks to represent the problem. How are you thinking about representing this problem?"
Name What Is Observed		"I notice that you have measured the length and width of the rectangle. Can you tell me a little more about what you are doing?"
Restate and Model Use of Mathematical Terms	Students who need support using mathematical vocabulary and expressing mathematical ideas	"So you used mental math to find the *product*." "Tell me how you found the *common denominator*."
Ask for Clarification	Students who need practice communicating mathematically	"Can you tell me how you know that these numbers are *prime numbers*?"

(Adapted from Anderson 2000, 97)

Strive to Understand and Respond to Students' Thinking

It is not an easy task to understand the mathematical thinking of students. Their words do not always convey their thoughts clearly. Yet we confer with students to "get a handle on what the student is doing (or trying to do)" (Fletcher and Portalupi, 2001, 50). So during math conferences, we take into consideration what we already know about our students from their past mathematical work, how they explain and reflect on their work, and their nonverbal cues.

Reference to anecdotal records allows teachers to refresh their memories about their students' mathematical histories. It takes more than one conference to gain an understanding of a student's approach to mathematics, what previous experiences he or she has had, and how to best respond to his or her learning needs. This kind of understanding develops over time. After several conferences, teachers have a better understanding of their students and can refer back to their conference notes. This allows them to begin their math conferences with students' strengths and weaknesses already in mind (Anderson 2000).

Probing, open-ended questions lead students to share their own mathematical thoughts. These questions not only prompt students to explain their work, but more importantly, to reflect on their mathematical thinking. Anderson (2000, 9) states, "[g]ood writers use strategies and techniques *thoughtfully* because they've learned to step back from their writing and reflect on what they're doing." So, too, do good mathematicians use strategies and techniques thoughtfully when they take time to reflect. Teachers may find, however, that students have trouble talking about their mathematical work or that what their students tell them does not really describe what the student has done or is doing. Sometimes, learners simply lose focus in the midst of describing their work. In these situations, it is vital for teachers to help them become more insightful and reflective.

As teachers, we want our students to become empowered mathematics learners. When learners are empowered, they assume ownership of their learning; believe in themselves as hard workers who embrace challenges; monitor their own thinking; recognize that confusion, errors, or difficulty are just part of the journey to becoming a proficient mathematician; and advocate for themselves through questioning to clarify meaning when necessary (Wedekind 2011). Reflection is an essential part of being an empowered mathematical learner.

According to Wedekind (2011, 169–170), when teachers explicitly notice the mathematical behavior and practices of their students, it encourages learners to be more reflective. Statements such as *I noticed you drew a diagram to help you solve the problem. Why did you choose to do that for this problem?* both make the mathematical thinking and actions of students more visible to them and lead students to reexamine their thinking. As a result, they may either revise their thinking or become better able to verbally express it. This kind of reflective thinking by students helps them to slow down and carefully analyze what they are doing. It also gives teachers much richer insight into their thinking.

In addition to considering what they already know about their students and listening attentively to their students during conferences, teachers should also be attuned to students' nonverbal cues. Body language and tone of voice may convey as much as the words spoken. If a student is listless and hesitant to speak about his or her work, it is important to find out why. Is the ennui math-related or due to other causes? Mistaken assumptions about the reasons for a student's languor or lack of engagement lead to teaching points that fail to target the needs of the student. The same is true about mistaken assumptions about students' enthusiasm for their work. In either situation, teachers should take care to understand their students' lines of thinking so their instructional responses are effective.

Once teachers have an understanding of a student's line of thinking, they have to decide whether to get behind that line of thinking during the conference or not. Teachers should make this decision by considering the student's immediate needs, if he or she is to grow mathematically. Teachers may choose to get behind the student's mathematical approach and teach him or her how to use it more effectively (Anderson 2000). Helping students expand their understanding or ability to apply a strategy they are using on their own most likely means that they will be more interested in the teaching point and more motivated to use what they learn later as they work. It makes an existing tool in their toolbox more efficient to use. Following the lead of students also shows a respect for them as mathematicians.

There is often a tension that teachers wrestle with when deciding between getting behind the thinking of their students or addressing other mathematical areas with which students may need help. In these situations, teachers have to weigh which will be of greatest value to the student. Sometimes, it is hard for teachers to let go of their own teaching agendas to follow the lead of their students, but that may be what is needed most. On the other hand, teachers must be willing to shift the focus of the conference away from the students' line of thinking if that is what will most effectively meet students' needs.

Build on Student Strengths and Interests

Whether teachers decide to get behind the thinking of their students or to address other learning needs, conferences are most meaningful when teachers build on the strengths and interests of the young mathematicians with whom they are conferring.

Graves (2003, 100) offers teachers wise advice when he writes: "The teacher looks for a child's potential in the words used in conference, the content of the piece, and the way the child goes about the craft. A teacher who looks for potential finds listening and observing an exciting venture." In the case of math conferences, one might substitute the words *math work* for *piece*. When teachers confer with young mathematicians looking for glimpses of their mathematical curiosity, knowledge, skills, and perseverance rather than solely for the mastery of a specific standard or use of a particular strategy, these mathematical conversations can indeed be astonishing.

Because we are so focused on the immediate instructional needs of our students, many of us discover that we are far better at locating errors and needs than we are at recognizing areas of strength or celebrating mathematical curiosity, attempts at experimentation, or interest in mathematics displayed by learners. Yet young mathematicians tend to be "fragile, highly sensitive, breakable creatures" (Fletcher and Portalupi 2001, 51). Drawing attention to even fledgling mathematical efforts leads them to a much more robust attitude toward the discipline of mathematics. Giving them concrete praise that notes specifically what they have done well makes them much more receptive to later conference teaching points and opens the door to further learning.

At the beginning of the school year when relationships between students and their teachers are just emerging, teachers should be especially cognizant of the fragility of students' egos. Even students who assume a confident air may mask their uncertainty about their mathematical abilities. When teachers meet with students for their first few math conferences, they strengthen their rapport with learners when their questions address areas of students' strength. That allows students to shine and gain confidence in their mathematical prowess.

As teachers learn more about their students' interests during conference conversations, they can differentiate their instruction by making their teaching points more relevant to students—not only the content of the teaching point, but also its context. If a student is a sports fan, the teaching point might relate the math content being taught to sports. For students who are interested in music, a teaching point might help them make connections between math and music. By tailoring instruction to help young mathematicians recognize the relevance of math to their own particular areas of interest, teachers create additional motivation for their students.

Teach Just One Thing

During a conference, "there is a natural flow that begins with understanding and moves toward teaching a particular skill, technique, or strategy. These one-on-one encounters are relatively rare in the bustle of classroom life, and they give you a rich opportunity to stretch the writer you're working with" (Fletcher and Portalupi 2001, 52). Although this statement refers to writing conferences, the same is true for Guided Math conferences. Math conferences provide opportunities to stretch young mathematicians. Because of the sporadic nature of conferring, it behooves us to use these opportunities wisely.

To make the most of conferences, teachers should teach just one thing—no more (Graves 2003). Teachers who are conferring with their students for the first time tend to over-teach. It is common for teachers to note several possible teaching points as they confer with a learner, but it is important to narrow it down to only one for two important reasons. First, teachers do not have the luxury of conducting lengthy math conferences. To use math conferences effectively, they must be brief. Otherwise, teachers would be unable to confer with enough students in their classes to make conferring worthwhile. Instead of conducting conferences, teachers would be essentially providing individualized instruction. Of course, there are not enough hours in a school day, nor enough teachers in each classroom, for that to occur. So it is imperative that math conferences be kept short and focused on only one teaching point.

The second reason for limiting the teaching within a math conference to just one point is to maximize its impact. Over-teaching frequently leaves students more confused about their mathematical work than they were before the conference. With the few minutes they have to confer with a student, a teacher can effectively share only a single teaching point.

When teachers identify several possible teaching points for a student, how do they decide which to pursue during a conference? The selection of a teaching point will be unique for each conference. Teachers have to rely on their professional judgment, drawing upon their knowledge of both the curriculum and the student. There are several factors that need to be considered when making this decision:

- **Does the teaching point address knowledge or skills that are prerequisite for the mathematics currently being taught?** If students have gaps in background knowledge or skills that must be addressed before they can master current mathematical content being taught, teaching points to fill these gaps are a priority.

- **Will the teaching point give the student additional mathematical understanding that will not only help him or her with the current work, but will give him or her a better understanding of mathematics for future mathematical work?** While it may be tempting to teach students *shortcuts* that will help them find solutions to problems they are working on at the moment, unless they understand the mathematics behind the shortcut, it will have little long-term benefit. Instead, teachers should focus on teaching points that truly lead to student growth in mathematical proficiency.

- **Does the teaching point align with current instructional focus?** This question does not refer to knowledge or skills that are prerequisite for the current focus of instruction, but instead refers directly to new concepts and skills being taught. Teachers may choose to place a greater importance on teaching points that support the mathematics being taught in either whole-class or small-group lessons.

- **Is a math conference the best setting for sharing the teaching point?** In some instances, identified needs may be addressed more productively in either whole-class or small-group settings. When this is the case, a teacher would be wise to share another teaching point during the conference, and then plan whole-class or small-group lessons for additional instruction.

- **Does the teaching point address a need the student has identified?** When students actively monitor their own comprehension and mathematical proficiency, their efforts should be lauded and respected. Teachers demonstrate respect for their students when they set aside their own teaching agenda at times to respond directly to self-identified student needs.

- **Is the teaching point related to the student's line of thinking?** Whenever possible, teachers should try to link the conference teaching point to the line of thinking the student has shared during the research phase of the conference. Students more readily recognize the value of the teaching point when they can see its relevance to their own mathematical thinking.

During a math conference, it is not unusual for several learning needs to be identified. Narrowing those down to just one teaching point may be difficult. Teachers' struggles over how to decide what is best for each student is inherent in the conferring process, but it may help teachers to know that there is rarely *one right* teaching point. If there is one, it will be obvious. Teachers should be confident that they know their students' mathematically better than anyone else—especially when they confer with them. If they draw upon what they know about their learners and about the mathematics curriculum to choose a conference teaching point, their students' mathematical understanding will be enriched.

Avoid Power Struggles

Experienced teachers know the importance of avoiding power struggles with students; they are simply no-win situations. This does not mean that teachers give up their authority, but instead they recognize the futility of engaging students in battles and find ways to avoid them. The concept of authority shifts from one of unilateral imposition to "taking the initiative to work with students toward a shared goal" (Hiebert et al. 1997, 40).

In math conferences, power struggles may arise when students have mathematical misconceptions or errors in their mathematical work. Student misconceptions and errors do need to be addressed. First, however, it is important to be sure that a misconception actually exists. Since young mathematicians are inexperienced in communicating mathematically, their explanations of what they are thinking may be poorly expressed and even misleading. Before attempting to correct what appears to be a misconception, teachers should use the conference research phase to find out if it really exists. Asking students to show what they are thinking using manipulatives or pictures makes their thinking more transparent. Teachers can also ask *why* questions to encourage learners to more clearly explain their ideas. If teachers verify that students really are working under misconceptions, those may be addressed as the teaching point of the conference.

Power struggles also often develop over mathematical errors. Rather than taking students' work and summarily marking the errors, teachers minimize mistake anxiety if they first point out any parts of the task that students may have done correctly before focusing on the errors (Willis 2010). When teachers are attuned to the likelihood of math stress in their students and take steps to ease it, they can use the teaching point more readily to help students to identify the causes of their errors and remedy them. As another method of avoiding power struggles over error corrections, teachers should work to create a classroom environment that teaches students the value of learning from mistakes. Ongoing discussions throughout the year about the importance of reflecting on errors to learn from them leads students to being more receptive to analyzing and learning from their errors (Willis 2010).

Another conference situation that may potentially bring about power struggles is when teachers require the use of a particular algorithm for computation, but students invent or generate their own procedures for computation instead. While students certainly need to acquire efficient methods to compute, research shows that having students invent algorithms leads to more flexible thinking, better understanding of place value, and more accurate solutions. Many mathematics educators suggest that teachers provide students with many opportunities "to develop, use, and discuss a variety of methods" for computation (Chapin and Johnson 2006, 44). When students develop their own procedures rather than just following those they were taught, they have to reflect on the mathematical meaning of their work (Hiebert et al. 1997)—something we want our students to do.

After a period of experimentation, standard and alternative algorithms can be introduced to "expand their repertoire of efficient, reliable, and generalizable methods" (Chapin and Johnson 2006, 44). Burns (2007, 187) advises teachers: "[I]nstruction should focus students on multiple strategies for computing and help them explore the parallels and differences among them." Math conferences are an ideal forum for teachers and students to compare invented, standard, and alternative computational methods and investigate the validity and utility of each. Such an approach reduces power struggles between students and teachers while at the same time enhancing student learning.

Set Clear Expectations for Students

Math conferences have a greater impact on future student mathematical work when teachers make it very clear what they want students to do after the conference. Explicitly sharing their expectations of follow-through makes it much more likely that learners will use what they have just learned from the conference teaching point. Jotting down a few ideas on a sticky note for students to refer to as they work is one way of making expectations clear. Another option for supporting student follow-through is asking students to make an "assignment box" in their math journals where they record how they will use what they learned (Anderson 2000, 66).

In addition to encouraging students to apply what was discussed, these written notes help students recall important aspects of the teaching points as they work. For students who often have trouble following through on the teaching point, teachers may choose to have them retell the conference during the link phase of the conference. At the conclusion of a conference, teachers can remind students that they will be back to see how the mathematical ideas just learned are being applied. Sometimes, students are encouraged to immediately begin using new ideas or strategies when teachers pause to observe students work before moving on.

There will be times when students fail to follow through with the work that was discussed during the conference. When that happens, teachers will want to conduct follow-up conferences asking students specifically why they have not followed through. If learners did not understand the teaching point, it can be retaught immediately—most often more successfully. After some experience trying to apply what was taught in the initial conference, students should have a better grasp of where their comprehension failed and be more attentive. Meeting again for follow-up conferences sends a clear message to learners that they are indeed expected to apply what they discussed during their conferences.

Teachers should be prepared for the few students who deliberately choose not to follow through. When this happens, these students should be made fully aware that this is unacceptable and has consequences. What those consequences are will vary from teacher to teacher, but should be consistent with those that result from not completing other assigned class work.

Encourage Deeper and More Complex Thinking

The comfortable conversational give-and-take between conferring students and teachers is ideal for prompting students to think more deeply about mathematics. When teachers share their own curiosity and wonder about math with their students, young learners begin to realize that mathematics is not a stolid subject with rigid rules learned from a book—but instead, a living and evolving discipline. They discover that "mathematicians are human beings who see their world through a mathematical lens" (Fosnot 2007, 9).

Identifying themselves as mathematicians seeking answers about the mathematical world around them—not just answers to assigned textbook problems—students' perspectives shift and broaden. Learners begin to wonder as they work and ask the "why" questions that lead them to more complex thinking. They enter a world very distinct from the traditional elementary school mathematics experience. Fosnot (2007, 9) vividly describes her transition from primary school mathematics to the real world of mathematics:

> *I don't remember really loving mathematics until I hit geometry in high school and was finally asked to actually do mathematics—to come up with my own proofs. Wow! Now that was fun, like cracking a mystery. And the more elegant the proof, the more beauty! When I reflect on my own experiences, which I know were similar to many other students', I often think how divorced primary school mathematics was from the real world of what mathematicians do. Mathematics is a highly creative activity. Mathematicians solve problems, but they also pose problems. They inquire. They explore relations, investigate interesting patterns, and craft proofs. They present their ideas to the mathematics community and those ideas hold up only when the logic of the argument is accepted. Real mathematicians don't line up before a wise one who checks their answers with a red pen!*

There is no reason why students should have to wait until they are in high school to "actually *do* mathematics." While not the only way to open students' eyes to math as a living discipline, math conference discussions offer rich opportunities for teachers to introduce students to the real world of mathematics.

In addition to students developing an understanding of the complexity of real world mathematics, educators know how important it is for them also to have opportunities to construct mathematical understanding. When presented with challenging problems, students apply strategies and procedures that are meaningful to them—drawing upon what they already know. From these experiences, they construct a more rigorous understanding of the subject and, building on that, begin to generalize—applying their insight to other contexts.

To get the most from this learning process, young mathematicians need sounding boards—someone who listens attentively to their ideas and with whom they can try out their thinking. Teachers play this role to a limited degree in either whole-class or small-group settings. But they can do this most effectively in more intimate conference conversations. When the ideas students express are unclear, probing questions can clarify their students' thinking and prompt even deeper, more complex thinking.

Celebrate Mathematical Growth

Much is revealed about students' mathematical growth and achievement during conferences. Mathematical progress becomes readily apparent to both teachers and young mathematicians as they talk. "Focusing on knowledge gain also provides a legitimate way to recognize and celebrate— as opposed to reward—success" (Marzano 2007, 27). Mathematical growth is definitely reason for celebration, and teachers and students should react with a sense of joy when it is discovered during conferences.

Celebration of their achievements helps students put what they have learned in perspective. They become more aware of their learning goals. With pride, they begin to notice even incremental gains in their mathematical knowledge and skills. And along with that pride comes increased desire to make further gains.

Beyond sharing a joyful sense of achievement with students during conferences, teachers may celebrate students' mathematical growth in the following ways (Anderson 2000):

- Ask the class to pause in their work. Then, describe the student's achievement for the class to share and celebrate.

- Call upon the student to share what he or she learned and how it impacts his or her mathematical work.

- Create a "Let's Celebrate" bulletin board. When mathematical growth is noted, the student records what he or she learned and posts it on the bulletin board.

- Share a student's success with his or her parents. Either send home a brief note describing the student's mathematical growth or telephone the parents that evening to share the good news.

Incorporate Humor and Playfulness

To many students—and teachers, too—mathematics is a deadly serious business. And in some respects, it is. Mink (2010, 7) states that "[m]athematics is one of the most feared subjects in school, yet it is a subject students will need for the rest of their lives." In today's world, mathematical literacy is vital. No longer does society believe that rigorous mathematics education is only for those who pursue highly specialized fields such as engineering or science (U.S. Department of Education 2008, 5). It is for all. So it is only fitting that it is taken seriously in the classroom.

Yet as any experienced teacher knows, humor and playfulness have a valuable function in the classroom—particularly when some students approach mathematics with trepidation, some with real anxiety. Humor can relieve the tension that students may have when conferring with teachers about mathematics. In fact, Graves (2003, 280) states: "If I could choose one common element that characterizes successful scaffolding conference work, it would be humor." The humor may come initially at the expense of the teacher. If teachers can laugh at themselves as they share their mathematical thoughts, it gives students permission to be more relaxed.

Rarely do students need to be reminded how important it is to learn math. No matter how irrelevant it may seem to them at times, there is general recognition among students that it is something they must master to be successful in later life. Humor in no way diminishes the value students place on their mathematical achievement. To the contrary, teachers who maintain a lighthearted demeanor as they discuss mathematics with students do much to inspire students to view mathematics more positively as a living discipline—albeit, one they can play with as they learn.

Use Conferences Strategically

Although math conferences are conducted with only one student at a time, teachers extend the instructional reach of conferring when they use them strategically. By conferring with individual students where they are working, the conversation is often overheard by the students sitting nearby as well. Knowing that eavesdropping occurs, teacher can strategically plan their conferences to indirectly share the teaching points with clusters of students (Anderson 2000; Serravallo and Goldberg 2007).

Why not simply meet with these students as a small group? To answer that question, it is important to reflect on the benefits of math conferences. Many of these benefits are the byproducts of having just one teacher and one student talking and sharing ideas, mathematician to mathematician. Moreover, instruction is more effective in a conference because the math concept or skill focused on in the teaching point is taught in the context of a students' present work rather than in isolation. As a result, it is more likely to become a part of the students' mathematical repertoire (Graves 2003). With a small-group format, those benefits are eliminated. Taking advantage of students' propensity to eavesdrop, however, allows the content of the conference to be imparted to several students while the intimate nature of the conference is retained.

There may be times when teachers will encourage neighboring students to stop their work and listen in. This instructional format is something of a hybrid between conferences and small-group lessons. The conference conversation itself still remains limited to one student and one teacher, but other students are formally included in the instruction as listeners. In such cases, there should be an expectation that they will apply what they have learned in their future mathematical work. It is especially necessary in these situations to follow-up with the listeners later to ensure their understanding of the teaching point.

Chapter Summary

The conference process may seem daunting to teachers at first. When teachers and students confer, their roles shift and blend—a disconcerting experience for them both. Teachers must be aware of what that shift in roles entails on their part. Instructional techniques they have relied on for other types of instruction may not prove as effective when conferring. Their first conferences especially demand their flexibility to develop rapport with students. Learners may also feel unsettled initially. It can be a slow process to teach students that their mathematical conversation is both valued and expected. But the beneficial outcomes of math conferences between teachers and students make this process worthwhile.

Individual conferences are rarely perfect. Teachers reflecting back on their conferences will always see things they wish they had done differently, but over time "change comes not because of one conference, one teacher, one topic, but a host of factors carefully orchestrated over time. There is no hurry, just the need for persistent observation and listening.... Since so little actual listening and specific response is usually given to learners or writers in any subject, time indeed is on the side of both the child and the teacher" (Graves 2003, 215–216).

REVIEW AND REFLECT

1. Which of these guidelines do you think is most important for effective teacher-student math conferences? Why? Are there any of these guidelines with which you disagree? If so, why?

2. Which of these strategies do you think will be most challenging for you when you confer with your students? Why?

3. If you already confer with your students about their reading and writing, how do math conferences compare to your literacy conferences? What can you take from your experience with conducting literacy conferences that will help you confer with students about their mathematical work?

AFTERWORD

As this book describes, the benefits of Guided Math conferences are many, both for teachers who are implementing the Guided Math framework and for those who are not. These one-on-one mathematical conversations between teachers and students enhance both the teaching and learning of mathematics. For teachers who are implementing Guided Math, however, they should be an integral part of their daily instruction. Not only are these conferences one of the seven components of the framework, but they also augment the effectiveness of the other components. Although some may consider small-group instruction the identifying feature of Guided Math (and it is does indeed provide one of the most valuable instructional tools of the framework), effective math conferences positively impact *all* aspects of the framework. Initially, teachers may be apprehensive about conferring one-on-one with their students. But with some experience, they learn the value of these robust mathematical conversations.

These intimate mathematical exchanges between students and teachers enhance the other components of Guided Math: creating a classroom environment of numeracy, math warm-ups, whole-group instruction, small-group instruction, math workshop, and assessment.

Creating a Classroom Environment of Numeracy

For a true classroom environment of numeracy to exist, it is important that students consider themselves valued members of a mathematical community of learners in which ideas are shared, problems are solved, skills are honed, and their understanding of math is deepened. With the use of math conferences, teachers encourage this sense of community by modeling how to discuss mathematical ideas with precision and by using appropriate mathematical vocabulary. They also demonstrate ways to respectfully disagree with the reasoning of others. Students, in turn, are able to practice discussing their own thinking in a nonthreatening and supportive setting. These experiences are valuable preparation for students' participation in Math Huddles (group discussions about mathematics) and in math talks with fellow students both in partners and in small groups.

Math Warm-Up

Math warm-ups set the mathematical tone for the day as students enter the classroom. To make the most of these brief tasks, teachers must take into account the learning needs of their students in light of the mathematics curriculum being taught. The knowledge gained by conferring with students often determines the choice of warm-up tasks so that they accurately target student needs.

Whole-Group Instruction

As with math warm-ups, whole-group instruction is most effective when it addresses students' needs. Guided Math conferences offer teachers a wealth of information about what those needs are. If the same gaps in understanding or skills are discovered during math conferences with many students, teachers may choose to focus on those gaps with whole-group lessons. Conferences may also spotlight math topics that require review in order for students to maintain their understanding and mastery. In addition, during conferences, some students may pose "I wonder…" questions or share particular interests about math that can later be addressed in whole-group instruction. These are often questions or interests that other students share but may not have expressed, so they are rich sources of ideas for stimulating student curiosity and motivation during whole-group mathematics lessons.

Small-Group Instruction

Perhaps the component that benefits most from ongoing teacher-student math conferences is small-group instruction. To effectively use this type of instruction, timely assessments are essential. Since math conferences serve as accurate and timely formative assessments, the information teachers obtain from these discussions with their students allows them to form groups of learners who have similar immediate instructional needs. Thus, they play a crucial role in guiding teachers as they flexibly and fluidly adjust the composition of their small groups. And with this insight into their students' degree of understanding and skill, the focus of the lessons can be differentiated to meet the unique needs of each small group.

Moreover, if teachers notice that several students are struggling to understand a math concept, they may decide to confer with a few of these learners to discover the root of their confusion or lack of understanding. This sampling of student thinking sometimes reveals underlying gaps in foundational knowledge that extend beyond the few students with whom the teacher conferred. Armed with this information, small-group lessons for these struggling students may fill those gaps and move these students to successful mastery of the current mathematics content being taught.

Math Workshop

Teachers using the Guided Math framework are acutely aware of the importance of the workstation tasks that students complete independently during Math Workshop. For Math Workshop to function well, the tasks must be both worthwhile and challenging. In addition, students must be able to complete them independently and with accuracy. Unless teachers are well acquainted with their students' capabilities, they may miss their mark as they plan these tasks. Tasks that are either too easy and regarded as "busy work" by students or that are too challenging so that students are unable to successfully complete them lead to Math Workshops that simply do not work. As a result, students may end up off-task, distracted, or frustrated. When this occurs, teachers also become frustrated and unable to use this time for small-group lessons or for math conferences because of frequent interruptions—either by questions from students or by the need to constantly manage the behavior of students working independently.

The better teachers know their students mathematically—what they are thinking and what their capabilities are—the more likely they are to be able to plan "just right" math workstation tasks that are appropriate for independent work by their students. Guided Math conferences offer teachers a manageable way to learn more about their students' mathematical thinking so they can fine tune the math workstation tasks to align precisely with their students' levels of proficiency.

Assessment

The entire Guided Math framework is contingent upon timely and specific assessment data—for planning lessons, planning math workstation tasks, and forming small groups for instruction. While benchmark tests, unit tests, and other written assessments may supply valuable information, they commonly lack timeliness. Too often, this information tends to be an assessment of what has been learned rather than an assessment tool for learning. Most assessments of this type give teachers little insight into students' mathematical thinking. Moreover, the most comprehensive assessment information is obtained when the assessment data is balanced, coming from observing students as they work, listening as they share their thinking, and considering their work products. Guided Math conferences allow teachers to conduct just such balanced assessments. These intimate conversations with students provide opportunities for all three aspects of balanced assessment.

Guided Math Framework

The Guided Math Menu of Instruction outlines how these components may be used in the classroom (Sammons 2010).

Guided Math Menu of Instruction
Daily: Classroom Environment of Numeracy Teachers create a classroom community where students are surrounded by mathematics. This includes real-life math tasks, data analysis, math word walls, instruments of measurement, mathematical communications, class-created math charts, graphic organizers, calendars, and evidence of problem solving.
Daily: Math Warm-Ups and Calendar Board This daily appetizer prepares the palate for the "Your Choice" entrees below with Math Stretches, Calendar Board activities, problems of the day, math-related classroom responsibilities, data work, incredible equations, reviews of skills to be maintained, and previews of skills to come.
Your Choice: Whole-Class Instruction This is an excellent teaching strategy to use when students are working at the same readiness level or when introducing lessons with a mini lesson or an activating strategy, teacher modeling and think-alouds, read-alouds of math-related literature, organizing a Math Huddle, reviewing previously mastered skills, setting the stage for Math Workshop, and using written assessments.
Your Choice: Small-Group Instruction Students are instructed in small groups whose composition changes based on their needs. The individualized preparation for these groups offers opportunities to introduce new concepts, practice new skills, work with manipulatives, provide intensive and targeted instruction to struggling learners, introduce activities that will later become part of Math Workshop, conduct informal assessments, and reteach based on student needs.
Your Choice: Math Workshop Students are provided with independent work to complete individually, in pairs, or in cooperative groups. The work may be follow-up from whole-group or small-group instruction, ongoing practice of previously mastered skills, investigations, math games, math journals, or interdisciplinary work.

Guided Math Menu of Instruction (cont.)
Daily: Individual Conferences To enhance learning, teachers confer individually with students, informally assessing their understanding, providing opportunities for one-on-one mathematical communication, and determining teaching points for individual students as well as for the class.
Daily: Assessment Be sure to include a generous helping of assessment *for* learning to inform instruction, with a dollop of assessment *of* learning to top off each unit.

Guided Math Conference Goals

Name _____   Date _____

Guided Math Conference Goals

Where I am:

Where I want to be:

What I will do:

When I will complete it:

How I will know when I am there:

Guided Math Conference Checklist

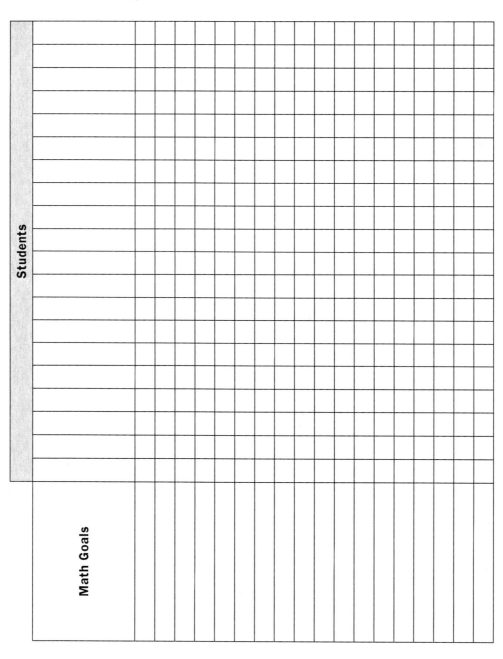

Students

Math Goals

Guided Math Conference Notes

Student	Date	Research	Compliment	Teaching Point

Sticky-Note Organizer

References Cited

Allington, Richard L. 2012. *What Really Matters for Struggling Readers: Designing Research-Based Programs. 3rd Ed.* Boston: Pearson.

Anderson, Carl. 2000. *How's It Going? A Practical Guide to Conferring with Student Writers.* Portsmouth, NH: Heinemann.

Andrade, Heidi L. 2010. "Students as the Definitive Source of Formative Assessment: Academic Self-Assessment and the Self-Regulation of Learning." *NERA Conference Proceedings 2010.* Accessed February 5, 2013. http://digitalcommons.uconn.edu/nera_2010/25.

Baldi, Stephane, Ying Jin, Melanie Skemer, Patricia J. Green, Deborah Herget, and Holly Xie. 2007. *Highlights from PISA 2006: Performance of U.S. 15 Year-Old Students in Science and Mathematics Literacy in an International Context.* Washington, DC: National Center for Education Statistics, Institute of Education Sciences, U.S. Department of Education. http://nces.ed.gov/pubs2008/2008016_1.pdf.

Black, Paul, and Dylan Wiliam. 2010. "Inside the Black Box: Raising Standards through Classroom Assessment." *Phi Delta Kappan* 92 (1): 81-90. http://www.questia.com/read/1G1-237749007.

Boushey, Gail, and Joan Moser. 2006. *The Daily Five: Fostering Literacy Independence in the Elementary Grades.* Portland, ME: Stenhouse Publishers.

Brookhart, Susan M. 2012. "Preventing Feedback Fizzle." *Educational Leadership* 70 (1): 24–29.

Burns, Marilyn. 2005. "Looking at How Students Reason." *Educational Leadership* 63(3): 26–31.

Burns, Marilyn. 2007. *About Teaching Mathematics: A K–8 Resource. 3rd Ed.* Sausalito, CA: Math Solutions.

Burns, Marilyn. 2010. "Snapshots of Student Misunderstandings." *Educational Leadership* 67 (6): 18–22.

Butler, Deborah L., and Philip H. Winne. 1995. "Feedback and Self-Regulated Learning: A Theoretical Synthesis." *Review of Educational Research* 65 (3): 245–281.

Butler, Ruth. 1988. "Enhancing and Undermining Intrinsic Motivation: The Effects of Task-Involving and Ego-Involving Evaluation on Interest and Performance." *British Journal of Educational Psychology* 58: 1–14.

Calkins, Lucy. 1994. *The Art of Teaching Writing.* Portsmouth, NH: Heinemann.

Calkins, Lucy. 2000. *The Art of Teaching Reading.* Boston: Allyn and Bacon.

Calkins, Lucy, Amanda Hartman, and Zoe White. 2005. *One to One: The Art of Conferring with Young Writers.* Portsmouth, NH: Heinemann.

Carr, Martha, and Barry Biddlecomb. 1998. "Metacognition in Mathematics from a Constructivist Perspective." In *Metacognition in Educational Theory and Practice*, edited by Douglas J. Hacker, John Dunlosky, and Arthur C. Graesser, 69–88. Mahwah, NJ: Lawrence Erlbaum Associates. http://www.questia.com/read/16186492.

Chapin, Suzanne H., and Art Johnson. 2006. *Math Matters: Understanding the Math You Teach, Grades K–8. 2nd Ed.* Sausalito, CA: Math Solutions.

Chapin, Suzanne H., Catherine O'Connor, and Nancy Canavan Anderson. 2003. *Classroom Discussions: Using Math Talk to Help Students Learn, Grades 1-6.* Sausalito, CA: Math Solutions.

Chappuis, Jan. 2009. *Seven Strategies of Assessment for Learning.* Princeton, NJ: Educational Testing Service.

———. 2012. "How Am I Doing?" *Educational Leadership* 70 (1): 36–41.

Christinson, Jan. 2012. "Integrating the Standards for Mathematical Practice with the Standards for Mathematical Content." In *Navigating the Mathematics Common Core State Standards*, 47–75. Englewood, CO: The Leadership and Learning Center.

Common Core State Standards Initiative. 2010. *Common Core State Standards for Mathematics*. Accessed January 30, 2013. http://www.corestandards.org/assets/CCSSI_Math%20Standards.pdf.

Conklin, Wendy. 2012. *Higher-Order Thinking Skills to Develop 21ˢᵗ Century Learners*. Huntington Beach, CA: Shell Education.

Davies, Anne. 2000. *Making Classroom Assessment Work*. Courtenay, Canada: Connections Publishing.

Davies, Anne, Caren Cameron, Colleen Politano, and Kathleen Gregory. 1992. *Together Is Better: Collaborative Assessment, Evaluation and Reporting*. Winnipeg, Canada: Peguis Publishers.

Dweck, Carol S. 2000. *Self-Theories: Their Role in Motivation, Personality, and Development*. Philadelphia: Psychology Press.

———. 2006. *Mindset: The New Psychology of Success*. New York: Random House.

Fisher, Douglas, and Nancy Frey. 2007. *Checking for Understanding: Formative Assessment Techniques for Your Classroom*. Alexandria, VA: Association for Supervision and Curriculum Development.

———. 2008. "Releasing Responsibility." *Educational Leadership* 66 (3): 32-37. Accessed February 4, 2013. http://www.ascd.org/publications/educational-leadership/nov08/vol66/num03/Releasing-Responsibility.aspx.

———. 2012. "Making Time for Feedback." *Educational Leadership* 70 (1): 42–46.

Fletcher, Ralph, and Joann Portalupi. 2001. *Writing Workshop: The Essential Guide*. Portsmouth, NH: Heinemann.

Fosnot, Catherine T. 2007. *Investigating Number Sense, Addition, and Subtraction, Grades K–3*. Portsmouth, NH: Heinemann.

Fountas, Irene C., and Gay Su Pinnell. 1996. *Guided Reading: Good First Teaching for All Children*. Portsmouth, NH: Heinemann.

———. 2001. *Guiding Readers and Writers, Grades 3–6*. Portsmouth, NH: Heinemann.

Graves, Donald. 2003. *Writing: Teachers and Children at Work*. Portsmouth, NH: Heinemann.

Gregory, Kathleen, Caren Cameron, and Anne Davies. 2011. *Knowing What Counts: Self-Assessment and Goal Setting. 2nd Ed*. Bloomington, IN: Solution Tree Press and Connections Publishing.

Grinder, Michael. 1995. *ENVoY: A Personal Guide to Classroom Management*. Portland, OR: Metamorphous Press.

Hattie, John. 2009. *Visible Learning: A Synthesis of Over 800 Meta-Analyses Relating to Achievement*. London: Routledge.

———. 2012a. "Know Thy Impact." *Educational Leadership* 70 (1): 18–23.

———. 2012b. *Visible Learning for Teachers: Maximizing Impact on Learning*. London: Routledge.

Hattie, John, and Helen Timperley. 2007. "The Power of Feedback." *Review of Educational Research* 77 (1): 81–112.

Hiebert, James, Thomas P. Carpenter, Elizabeth Fennema, Karen C. Fuson, Diane Wearne, Hanlie Murray, Alwyn Oliver, and Piet Human. 1997. *Making Sense: Teaching and Learning Mathematics with Understanding*. Portsmouth, NH: Heinemann.

Hoffer, Wendy W. 2012. *Minds on Mathematics: Using Math Workshop to Develop Deep Understanding in Grades 4–8*. Portsmouth, NH: Heinemann.

Hutchby, Ian, and Robin Wooffitt. 2008. *Conversation Analysis. 2nd Ed.* Malden, MA: Polity Press.

Irvin, Matthew. J., Judith L. Meece, Soo-yong Byun, Thomas W. Farmer, and Bryan C. Hutchins. 2011. "Relationship of School Context to Rural Youth's Educational Achievement and Aspirations." *Journal of Youth and Adolescence,* 40 (9): 1225–1242.

Jensen, Eric. 2013. "How Poverty Affects Classroom Engagement." *Educational Leadership* 70 (8): 24–30.

Johnston, Peter H. 2004. *Choice Words: How Our Language Affects Children's Learning.* Portland, ME: Stenhouse Publishers.

Kluger, Avraham N., and Angelo DeNisi. 1996. "The Effects of Feedback Interventions on Performance: A Historical Review, a Meta-Analysis, and a Preliminary Feedback Intervention Theory." *Psychological Bulletin* 119 (2): 254–284.

Marzano, Robert J. 2007. *The Art and Science of Teaching: A Comprehensive Framework for Effective Instruction.* Alexandria, VA: Association for Supervision and Curriculum Development.

Miller, Debbie. 2008. *Teaching with Intention: Defining Beliefs, Aligning Practice, Taking Action, K–5.* Portland, ME: Stenhouse Publishers.

Mind Tools. "Active Listening: Hear What People Are Really Saying." Accessed June 6, 2013. http://www.mindtools.com/CommSkll/ ActiveListening.htm.

Mink, Deborah V. 2010. *Strategies for Teaching Mathematics.* Huntington Beach, CA: Shell Education.

Moon, Jean, and Linda Schulman. 1995. *Finding the Connections: Linking Assessment, Instruction, and Curriculum in Elementary Mathematics.* Portsmouth, NH: Heinemann.

Murray, Donald M. 2004. *A Writer Teaches Writing Revised. 2nd Ed.* Boston: Heinle Cengage Learning.

National Center for Education Statistics. Institute of Education Sciences. U.S. Department of Education. "Trends in International Mathematics and Science Study." Accessed September 4, 2010. http://nces.ed.gov/timss/results07_math07.asp.

National Council of Teachers of Mathematics. 1991. *Mathematics Assessment: Myths, Models, Good Questions, and Practical Suggestions*. Reston, VA: National Council of Teachers of Mathematics.

———. 2000. *Principles and Standards for School Mathematics: An Overview*. Reston, VA: National Council of Teachers of Mathematics.

National Research Council. 2001. *Adding It Up: Helping Children Learn Mathematics*. Edited by Jeremy Kilpatrick, Jane Swafford, and Bradford Findell. Mathematics Learning Study Committee, Center for Education, Division of Behavioral and Social Sciences and Education. Washington, DC: National Academy Press.

Nicol, David J., and Debra Macfarlane-Dick. 2006. "Formative Assessment and Self-Regulated Learning: A Model and Seven Principles of Good Feedback Practice." *Studies in Higher Education* 31 (2): 199–218.

Nuthall, Graham. 2005. "The Cultural Myths and Realities of Classroom Teaching and Learning: A Personal Journey." *Teachers College Record* 107 (5): 895–934. Accessed January 24, 2013. http://www.tcrecord.org/content.asp?contentid=11844.

Polya, George. 1957. *How to Solve It: A New Aspect of Mathematical Method. 2nd Ed*. Princeton, NJ: Princeton University Press.

Sadler, D. Royce. 1989. "Formative Assessment and the Design of Instructional Systems." *Instructional Science* 18 (2): 119–144.

Sammons, Laney. 2010. *Guided Math: A Framework for Mathematics Instruction*. Huntington Beach, CA: Shell Education.

———. 2012. *Strategies for Implementing Guided Math*. Huntington Beach, CA: Shell Education.

Saskatchewan Ministry of Education. 2009. "Mathematics 3." Accessed January 25, 2013. www.education.gov.sk.ca.

Schunk, Dale H. 2003. "Self-Efficacy for Reading and Writing: Influence of Modeling, Goal Setting, and Self-Evaluation." *Reading and Writing Quarterly* 19 (2): 159–172.

Scriven, Michael, and Richard Paul. "Defining Critical Thinking." The Foundation for Critical Thinking. Accessed January 30, 2013. http://www.criticalthinking.org/pages/defining-critical-thinking/410.

Seeman, Howard. 1994. *Preventing Classroom Discipline Problems: A Classroom Management Handbook.* Lancaster, PA: Technomic Books. Quoted in Anderson 2000, 172.

Serravallo, Jennifer, and Gravity Goldberg. 2007. *Conferring with Readers: Supporting Each Student's Growth and Independence.* Portsmouth, NH: Heinemann.

Slaughter, Holly. 2009. *Small-Group Writing Conferences, K–5: How to Use Your Instructional Time More Efficiently.* Portsmouth, NH: Heinemann.

Steen, Lynn A. 1990. "Numeracy." *Daedalus* 119 (2): 211–231. Accessed January 22, 2013. http://www.stolaf.edu/people/steen/Papers/numeracy.html.

Stiggins, Richard J. 1997. *Student-Centered Classroom Assessment.* Upper Saddle River, NJ: Merrill/Prentice Hall.

———. 2002. "Assessment Crisis: The Absence of Assessment for Learning." *Phi Delta Kappan* 83 (10): 758–765.

Stiggins, Rick. 2005. "From Formative Assessment to Assessment for Learning: A Path to Success in Standards-Based Schools." *Phi Delta Kappan* 87 (4): 324–328.

———. 2007. "Assessment Through the Student's Eyes." *Educational Leadership* 64 (8): 22-26. Accessed February 4, 2013. http://www.ascd.org/publications/educational-leadership/may07/vol64/num08/Assessment-Through-the-Student's-Eyes.aspx.

Teacher Learning Collaborative. "Math Conferencing: Building Language Development and Math Competence with 4th and 5th Grade English Learners through the Use of Math Conferences." Antioch University Los Angeles. Accessed January 30, 2013. http://www.antiochla.edu/tlc/for-k-5-teachers/math-conferencing/math-conferencing.html.

Tomlinson, Carol A. 2013. "One to Grow On: Teachers Who Stare Down Poverty." *Educational Leadership* 70 (8): 88–89.

U.S. Department of Education. 2008. *The Final Report of the National Mathematics Advisory Panel.* National Mathematics Advisory Panel. http://www.ed.gov/MathPanel.

Van de Walle, John A., Karen S. Karp, and Jennifer M. Bay-Williams. 2010. *Elementary and Middle School Mathematics: Teaching Developmentally.* 7th Ed. Boston, MA: Allyn and Bacon.

Vygotsky, Lev S. 1978. *Mind in Society: The Development of Higher Psychological Processes.* Cambridge, MA: Harvard University Press.

Wedekind, Kassia O. 2011. *Math Exchanges: Guiding Young Mathematicians in Small-Group Meetings.* Portland, ME: Stenhouse Publishers.

Wiggins, Grant. 2012. "Seven Keys to Effective Feedback." *Educational Leadership* 70 (1): 10–16.

Wiliam, Dylan. 2012. "Feedback: Part of a System." *Educational Leadership* 70 (1): 30–34.

Willis, Judy. 2010. *Learning to Love Math: Teaching Strategies That Change Student Attitudes and Get Results.* Alexandria, VA: Association for Supervision and Curriculum Development.

Wormeli, Rick. 2006. *Fair Isn't Always Equal: Assessing and Grading in the Differentiated Classroom.* Portland, ME: Stenhouse Publishers.